ACE BIOCHEMISTRY!

(THE EASY GUIDE TO ACE BIOCHEMISTRY)

BY: DR. HOLDEN HEMSWORTH

DISCLAIMER

Biochemistry, like any field of science, is continuously changing and new information continues to be discovered. The author and publisher have reviewed all information in this book with resources believed to be reliable and accurate and have made every effort to provide information that is up to date and correct at the time of publication. Despite our best efforts we cannot guarantee that the information contained herein is complete or fully accurate due to the possibility of the discovery of contradictory information in the future and any human error on part of the author, publisher, and any other party involved in the production of this work. The author, publisher, and all other parties involved in this work disclaim all responsibility from any errors contained within this work and from any results that arise from the use of this information. Readers are encouraged to check all information in this book with institutional guidelines, other sources, and up to date information.

MCAT® is a registered trademark of the Association of American Medical Colleges and holds no affiliation with this book.

The information contained in this book is provided for general information purposes only and does not constitute medical, legal or other professional advice on any subject matter. The author or publisher of this book does not accept any responsibility for any loss which may arise from reliance on information contained within this book or on any associated websites or blogs.

WHY I CREATED THIS STUDY GUIDE

I love teaching biochemistry to students on a daily basis. From talking to students, I learned that biochemistry is considered one of the most difficult classes for undergrads and that it contains a lot of material that shows up on the MCAT. Because of this, it was my goal to create a book that could be a supplement to lecture notes and textbooks. I also wrote it with the hope that it could aid students in studying for classroom exams and also aid students that were studying for the MCAT.

In this book, I try to breakdown the content covered in most introductory biochemistry courses in college for easy understanding and to point out the most important subject matter that students are likely to encounter. This book is meant to act as a supplement to boost your learning, go hand in hand with your studying, and help increase your test scores!

Best regards,

Dr. Holden Hemsworth

TABLE OF CONTENTS

CHAPTER 1: Biochemistry – The Basics 1

CHAPTER 2: Aqueous Chemistry 5

CHAPTER 3: The Structure of Proteins 11

CHAPTER 4: The Non-catalytic Functions of Proteins 19

CHAPTER 5: The Catalytic Functions of Proteins... 28

CHAPTER 6: Enzyme Kinetics and Inhibition 33

CHAPTER 7: Lipids and Membranes 41

CHAPTER 8: Membrane Transport 48

CHAPTER 9: Signaling.. 56

CHAPTER 10: Carbohydrates 61

CHAPTER 11: Metabolism and Bioenergetics 67

CHAPTER 12: Glucose Metabolism 73

CHAPTER 13: The Citric Acid Cycle 85

CHAPTER 14: Oxidative Phosphorylation 90

CHAPTER 15: Photosynthesis 98

CHAPTER 16: Lipid Metabolism... 107

CHAPTER 17: Nitrogen Metabolism 118

CHAPTER 18: Regulation of Mammalian Fuel Metabolism 128

CHAPTER 19: Genes to Proteins 137

CHAPTER 20: DNA Replication and Repair 145

CHAPTER 21: Transcription and RNA... 158

CHAPTER 22: Translation 169

CHAPTER 1: BIOCHEMISTRY – THE BASICS

The Four Major Biomolecules

Biomolecules consist of carbon and hydrogen and are present in living organisms.

Carbohydrates

- Also called saccharides

- Contain multiple -OH groups

- Readily convertible between open and close (cyclic) form

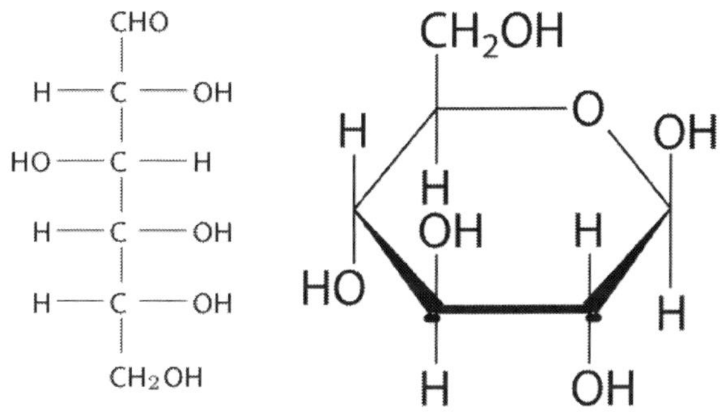

Carbohydrate Representation Examples

Nucleotides

- 5-Carbon Sugar

- Nitrogen containing base ring

- PO_3 group

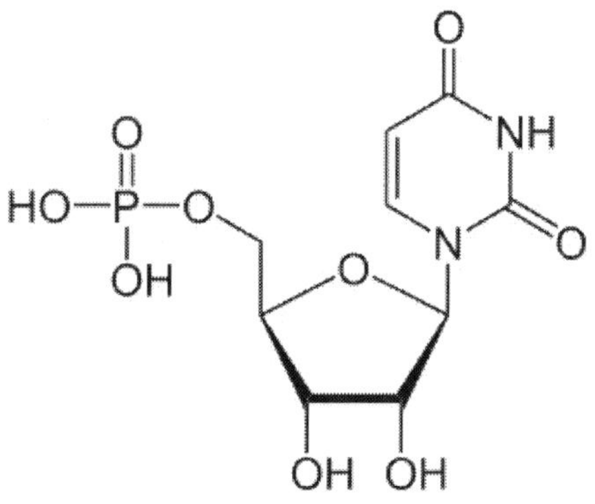

Nucleotide Representation Example

Lipids

- Long chain of hydrocarbons

- Mostly nonpolar

  - Generally, this makes them water-insoluble

- They can be amphipathic

  - Amphipathic - contain both polar and nonpolar components

- Steroids are lipids that have fused rings in their structure

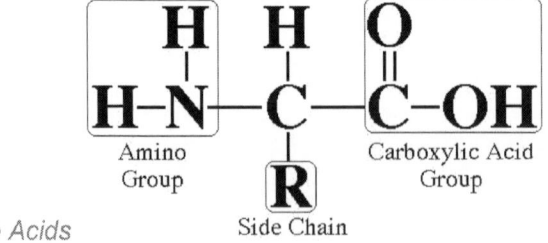

Cholesterol: *Triglyceride:*

Amino Acids

- All have at least 2 ionizable groups (an amino and a carboxyl group)

- Their identities are determined by the side chain (R group)

General Structure of Amino Acids

<u>Biological Polymers</u>

Polymers are built from unit molecules (monomers) for structure/nutrient storage. The monomers in polymers are called "residues."

Polysaccharides (Polymer of Carbohydrates)

- Form glycosidic bonds

- Structure varies from linear to highly branched

Proteins (Polymer Made of Amino Acids)

- Sometimes also called polypeptides

- Peptide (amide) bonds are formed between amino acid molecules to hold them together

- Wide variety of combinations are possible

Nucleic Acids (Polymers of Nucleotides - Polynucleotides)

- Both DNA and RNA are nucleic acids

- They form phosphodiester bonds

Lipids

- Lipids do not have a standard functional group

  - So they aggregate differently than amino acids, carbohydrates, and nucleotides

Gibbs Free Energy

Gibbs free energy can be used to determine the direction of the chemical reaction under given conditions.

- $\Delta G = \Delta H - T\Delta S$ or $\Delta G = G_{products} - G_{reactants}$

  - G = Gibbs free energy (J/mol)

  - H = enthalpy (J/mol) - total energy content of a system

  - S = entropy (J/K*mol) - measure of disorder or randomness (how energy is dispersed)

  - T = Temperature (K)

    - As T increases so does S

- $+\Delta G$ means energy must be put into the system

  - Indicates that a process is **nonspontaneous or endergonic**

- $-\Delta G$ means energy is released by the system

  - Indicates that a process is **spontaneous or exergonic**

- $\Delta G = 0$ indicates that the system is at equilibrium

Important: ΔG only indicates if a process occurs spontaneously or not, but does **not** indicate anything about how fast a process occurs.

The Three Domains of Life

Introduced by Carl Woese in 1977, the three domains of life is a system of biological classification based off of rRNA sequences of organisms.

- Bacteria

- Archaea

- Eukarya (eukaryota)

Prokaryotes and Eukaryotes

The bacteria and archaea domains consist of prokaryotes, while eukarya consists of eukaryotes.

- **Eukaryotes**

    o Have a true nucleus

    o Mostly comprised of multi-cellular organisms but some are unicellular

- **Prokaryotes**

    o Have no true nucleus

    o Have a nucleoid

    o Bacteria and archaea

    o Single cell

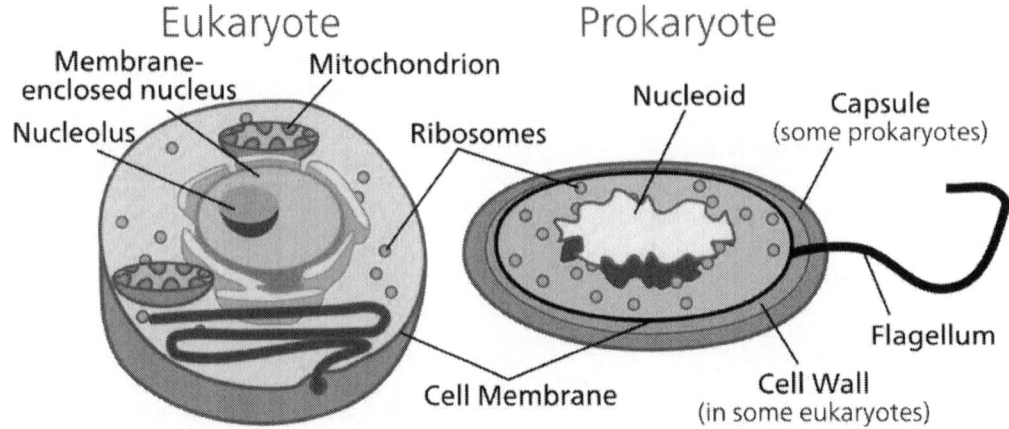

CHAPTER 2: AQUEOUS CHEMISTRY

Atomic Orbitals

Mathematical function that describes the wave-like behavior of an electron in a molecule, it calculates the probability of where you might find an electron (e⁻).

Orbitals can combine to form hybridized orbitals needed for molecular bonding interactions.

Types of Bonds

Covalent Bonds

- Two atoms share valence electrons

- Indicates that atomic orbitals are overlapping

    o Overlapping requires proximity and orientation

- Two Types

    o Non-polar covalent bond – electrons shared equally between atoms

        ▪ Electronegativity of the two atoms is about the same

        ▪ Typically electronegativity difference between the two atoms has to be less than 0.5 for non-polar bonds

        ▪ Electronegativity – an atom's ability to attract and hold on to electrons, represented by a number

    o Polar covalent bonds – electrons shared disproportionately between atoms

        ▪ Electronegativity between the two atoms is different by a greater degree than 0.5 but less than 2.0

Ionic Bonds

- Electrons are transferred, not shared between atoms

- An atom with high electronegativity will take an electron from an atom with low electronegativity

    o Typically, difference in electronegativity is more than 2.0

Hydrogen Bonds

- Attractive force between a hydrogen attached to an electronegative atom of one molecule to a hydrogen attached to an electronegative atom of a different molecule

- Electronegative atom is usually an O, N, or F

Van der Waals Forces

A general term used for the attraction of intermolecular forces between molecules.

Dipole-dipole interactions

- Interaction between 2 polar groups

- Permanent dipole moments

London dispersion forces

- Interaction between 2 non-polar molecules

- Small fluctuation in electronic distribution

Intermolecular Forces

Forces that act between neighboring particles (can be repulsive or attractive).

- Intermolecular bond strength ranking (strong to weak):

    o Covalent > ionic > hydrogen > Van der Waals forces

- Weaker bonds and forces are easily broken or overcome and also re-formed, making them vital for the molecular dynamics of life

Van der Waals Radius

The Van der Waals Radius is the distance from the nucleus to the electron cloud surface.

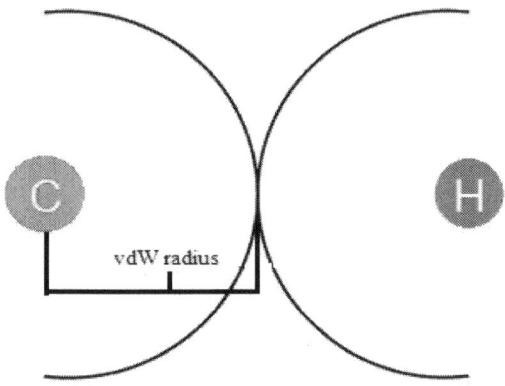

- When orbitals of two atoms are as close as possible but do not overlap, no bonds can be formed

  o This is shown in the picture above

- When atoms are close enough that their orbitals overlap, bonds can be formed

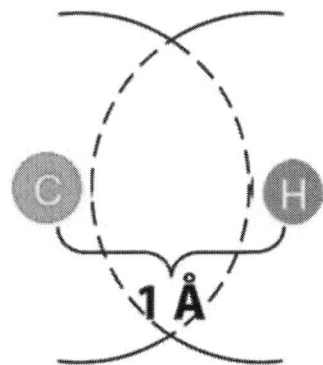

  o A covalent bond distance is about 1 angstrom (1 X10^{-10} meters)

  o A hydrogen bond distance is about 1.8 angstrom

Note: Shorter bond = stronger bond

Properties of Water

One of the most important properties of water is that it is less dense in solid form. This is because it expands as it solidifies, forming a crystalline structure maintained by hydrogen bonding.

- Highly cohesive

    o Because of hydrogen-bonding interactions

    o One molecule of H_2O can hydrogen bond with 4 other water molecules

- High surface tension

    o Tension on the surface of a liquid caused by attraction of particles in the surface layer that allows it to resist an external force

Molecular Bonds of Water

The electronic structure of water is tetrahedral (2 covalent bonds with H atoms and 2 sets of unpaired electrons).

- Water has 2 hydrogen atoms **covalently** bonded to 1 oxygen atom

- Water is a **polar** molecule

    o Polar atoms have dipoles because of unequal sharing of electrons

    o Dipoles can align to form H bonds

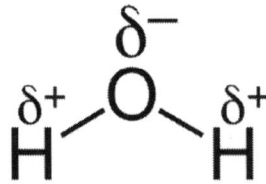

- H_2O can hydrogen bond with other molecules besides itself

    o Common electronegative atoms are N, O, and S

Hydrophobic Effect and Amphiphilic Molecules

The hydrophobic effect is the tendency of nonpolar substances to aggregate in aqueous solution and exclude water molecules. Amphiphilic molecules have both hydrophobic and hydrophilic parts.

- Hydrophobic – water "hating"

- In contrast, hydrophilic – water "loving"

Amphiphilic Molecules in Water

- Nonpolar tails (hydrophobic portion) point away from water

- Polar heads (hydrophilic portion) are exposed to water

- Different amphipathic molecules aggregate in different ways based on the number of tails and size of the polar head group

- Micelles form when there is only 1 tail and take spherical form in aqueous solutions

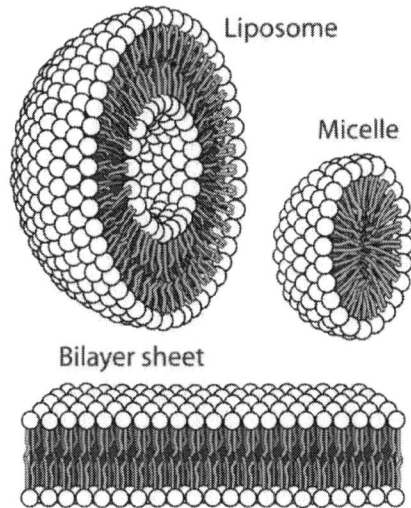

Acid-Base Chemistry

According to the Bronsted-Lowery definition an acid is proton donating and a base is proton accepting.

Henderson-Hasselbalch Equation

$$K_a = \frac{\left[A^-\right]\left[H^+\right]}{\left[HA\right]}$$

- K_a is the acid dissociation constant

- $pK = -\log K_a$

    - More acidic a compound, the larger the value of K_a

    - Larger K_a value equates to a smaller pK value

    - As pK decreases, the acid strength increases (greater tendency to donate a H^+)

pH Scale

- pH = -log[H⁺]   or $$pH = pK + \log \frac{[A^-]}{[HA]}$$

    ○ As a solution gets more acidic, the pH decreases+
- Many groups on biological molecules can donate or accept H^+

    ○ When the pH= pK, it means that ½ of the time the group in a solution is deprotonated (has donated its H^+)

    ○ When the pH < pK, it means that the group is protonated

    ○ When the pH > pK, it means that the group is deprotonated

Aqueous Buffers

Buffer solutions resist changes in pH when small amounts of acid or base are added to it.

Adding Acids to Bases

- Weak acids only partially dissociate in solution

- Strong acids will almost completely dissociate in solution

- Adding a strong acid to a weak base results in less of a pH change

    ○ Some of the acid is neutralized

- Adding a strong base to a weak acid results in less of a pH change

    ○ Some of the base is neutralized

Buffers Resist pH Changes

- Effective buffering range is pH = pK -1 to pH = pK +1

Things to Notice from the Titration Curve

- Rapid change in pH at the start and end points

- There is a much more gradual change in the effective buffering range

- Shaded area is the effective buffering range and varies by ~1 pH unit on either side of the midpoint (pK)

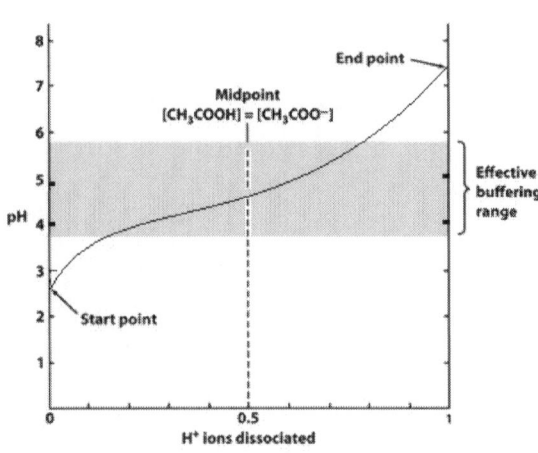

CHAPTER 3: THE STRUCTURE OF PROTEINS

Peptides, Polypeptides, and Proteins:

- A peptide is made up of two or more amino acids

- A polypeptide is a chain of many amino acids

- Proteins contain one or more polypeptides

    o Are long chains of amino acids held together by peptide bonds

The 20 Standard Amino Acids

Each amino acid has a full name (e.g., glycine), a 3 letter code (e.g., gly), and a 1 letter code (e.g., g) that identifies it.

The Hydrophobic Amino Acids – Typically Found Within the Protein Core

- Glycine (Gly, G)

    o Simplest amino acid (side chain is a single H atom)

$$COO^-$$
$$H-C-H$$
$$NH_3^+$$

Glycine Structure:

- Alanine (Ala, A) and Phenylalanine (Phe, F)

$$COO^-$$
$$H-C-CH_3$$
$$NH_3^+$$

Alanine:

$$COO^-$$
$$H-C-CH_2-$$
$$NH_3^+$$

Phenylalanine

    o Alanine has a methyl group (CH_3) for its side chain

    o **Phenyl**alanine has a **phenyl** ring on the alanine residue

- Valine (Val, V), Leucine (Leu, L), and Isoleucine (Ile, I)

Valine:

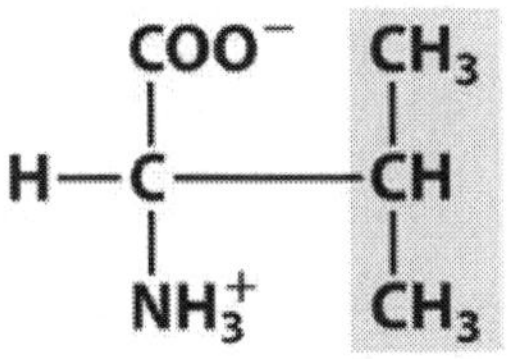

Leucine:

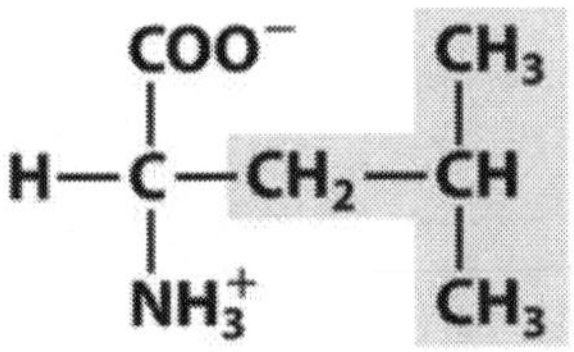

Isoleucine:

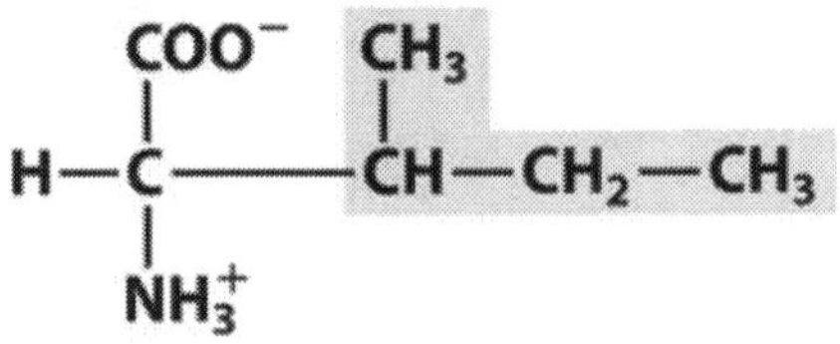

- o Leucine has an extra methylene group (CH_2) than Valine

- o Isoleucine has the same functional parts in its side chain as leucine

  - Just arranged differently

- Proline (Pro, P)

Proline:

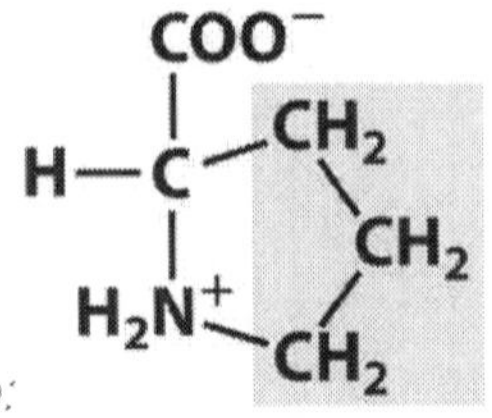

- o Proline is the only amino acid whose side chain loops back onto its own backbone

- o Proline produces kinks in the polypeptide sequence that it is a part of

The Polar Amino Acids – May Participate in Hydrogen Bonds

- Methionine (Met, M) and Cysteine (Cys, C)

Methionine

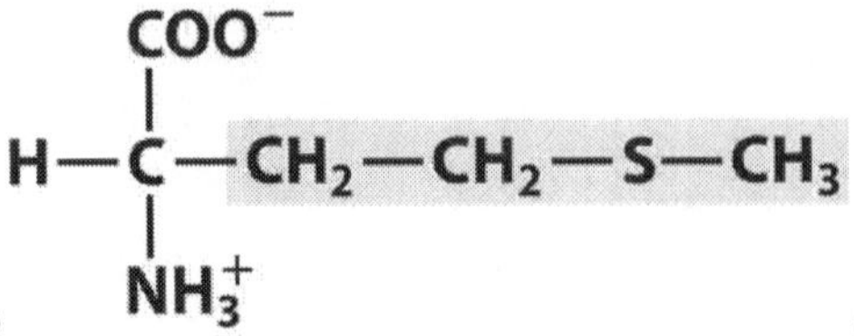

Cysteine:

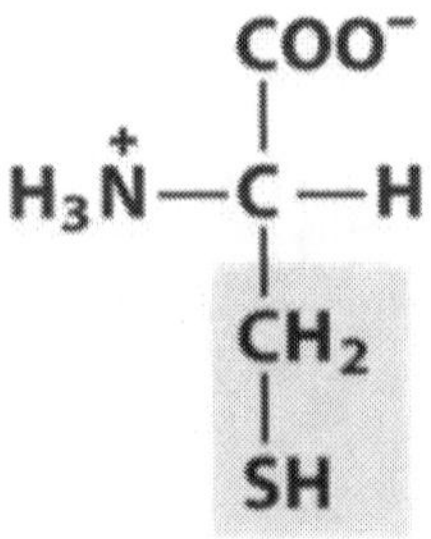

- o Methionine and Cysteine are the only two amino acids with a sulfur in their R-group

- Tryptophan (Trp, W)

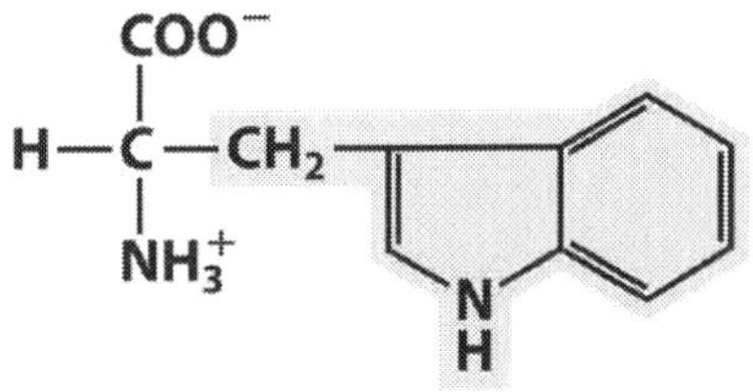

Tryptophan:

  - Tryptophan is the only amino acid with a fused ring system

- Serine (Ser, S) and Threonine (Thr, T)

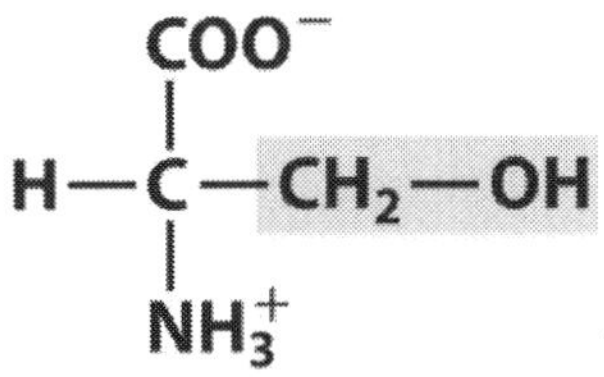

Serine: 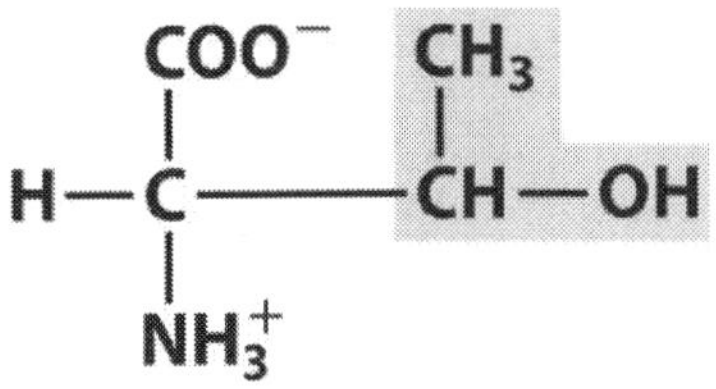 *Threonine:*

  - -OH groups are important nucleophiles in biochemical reactions

  - Other chemical groups can covalently bond to proteins via –OH groups

- Tyrosine (Tyr, Y)

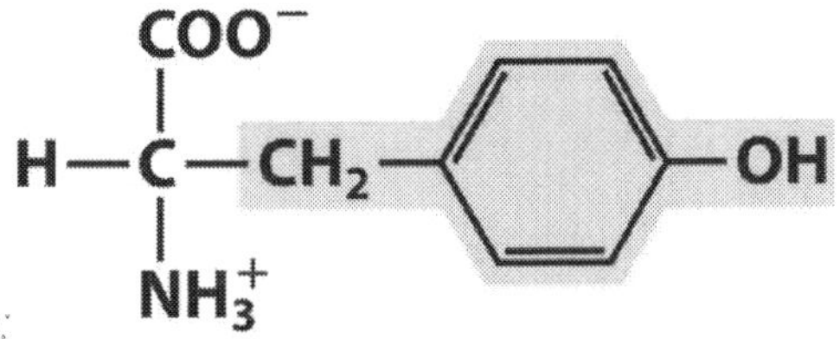

Tyrosine:

  - Tyrosine is a derivative of the nonpolar amino acid phenylalanine

- Asparagine (Asn, N) and Glutamine (Gln, Q)

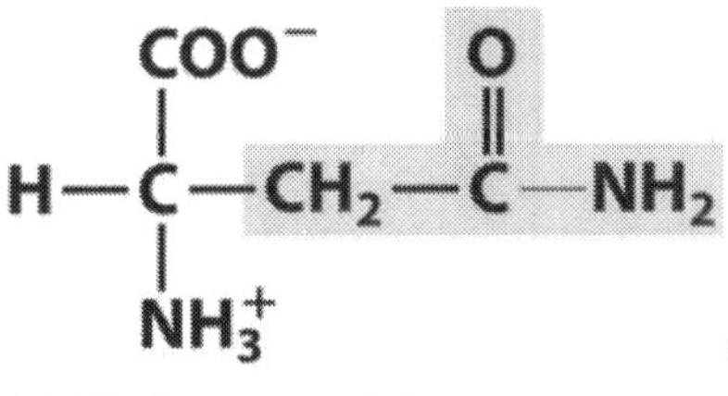

Asparagine: 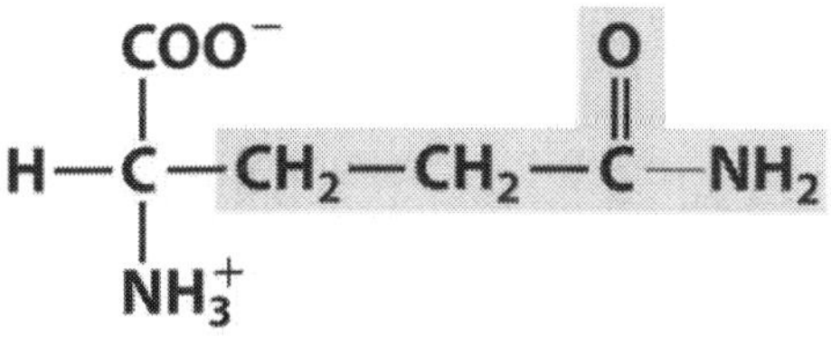 *Glutamine:*

  - Both have amide groups

  - Glutamine has 2 methylene groups, asparagine has only 1

- Histidine (His, H)

$$H-\underset{\underset{NH_3^+}{|}}{\overset{\overset{COO^-}{|}}{C}}-CH_2-\text{(imidazole ring)}$$

Histidine:

  - o Histidine has a 5-membered ring system with nitrogen atoms

  - o Histidine can be charged at certain pH values

  - o Histidine is a basic amino acid (capable of accepting H^+)

The Charged Amino Acids

- Aspartate (Asp, D) and Glutamate (Glu, E) – Carry a Negative Charge

$$H-\underset{\underset{NH_3^+}{|}}{\overset{\overset{COO^-}{|}}{C}}-CH_2-\overset{\overset{O}{\|}}{C}-O^-$$

Aspartate:

$$H-\underset{\underset{NH_3^+}{|}}{\overset{\overset{COO^-}{|}}{C}}-CH_2-CH_2-\overset{\overset{O}{\|}}{C}-O^-$$

Glutamate:

  - o Aspartate and glutamate are comparable to asparagine and glutamine

    - ▪ Except that aspartate and glutamate have carboxylate groups

  - o They are acidic amino acids (have H^+ to donate)

    - ▪ They have low pK values which mean they are usually deprotonated and carry a negative charge

- Lysine (Lys, K) and Arginine (Arg, R) – Carry a Positive Charge

$$H-\underset{\underset{NH_3^+}{|}}{\overset{\overset{COO^-}{|}}{C}}-CH_2-CH_2-CH_2-CH_2-NH_3^+$$

Lysine:

$$H-\underset{\underset{NH_3^+}{|}}{\overset{\overset{COO^-}{|}}{C}}-CH_2-CH_2-CH_2-NH-\overset{\overset{NH_2}{|}}{C}=NH_2^+$$

Arginine:

  - o Lysine and arginine have long side chains with a positively charged amine group

  - o They are basic amino acids (can accept H^+)

  - o They have high pK values

    - ▪ Usually protonated and carry a positive charge

The Four Levels of Protein Structure

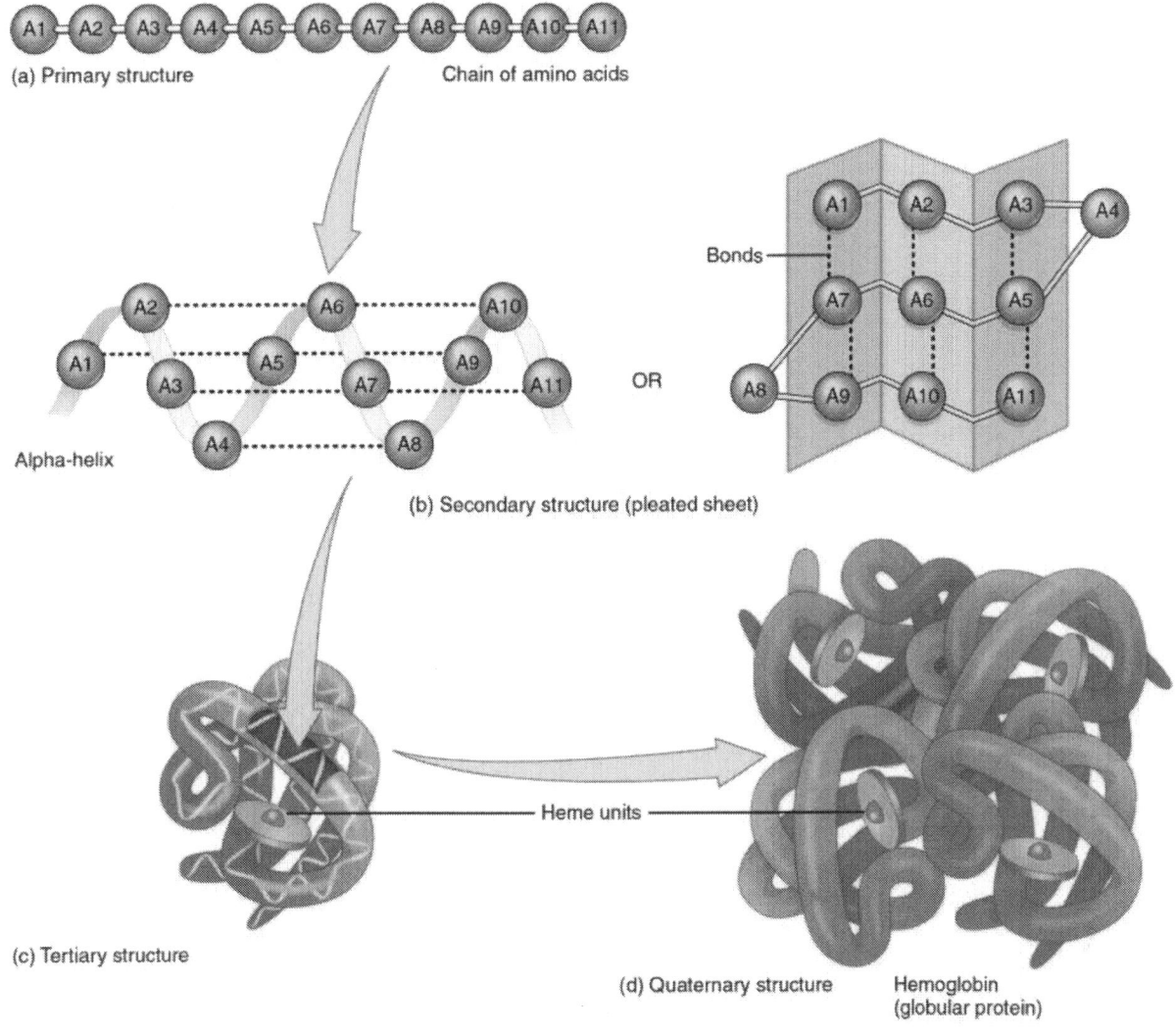

(a) Primary structure   Chain of amino acids

OR

Bonds

(b) Secondary structure (pleated sheet)

Alpha-helix

(c) Tertiary structure

Heme units

(d) Quaternary structure   Hemoglobin (globular protein)

Primary Structure

- Sequence of amino acids
- Bonds present: peptide bonds

Secondary Structure

- Localized conformation of the polypeptide backbone
- Bonds present: peptide bonds

- Two types of secondary structure:
    - Alpha Helix
        - Peptide backbone wound around a long axis core
        - Forms a cylinder
        - R-groups radiate outward
        - 3.6 amino acids per 360° turn
        - Formed by hydrogen bonds
    - Beta Sheet
        - A linear zig-zag sheet of polypeptides
        - Formed by hydrogen bonds, as well as intra and inter-chain reactions

Tertiary Structure

- 3D structure of an entire polypeptide
- Bonds present: hydrogen bonds, ionic bonds, disulfide bonds, and Van der Waals forces

Quaternary Structure

- Spatial arrangement of polypeptide chains in a protein with **multiple subunits**
    - May have prefix homo- (identical subunits) or hetero- (different subunits)
- Bonds present: hydrogen bonds, ionic bonds, disulfide bonds, and Van der Waals forces

Protein Folding

The hydrophobic effect has the most to do with protein folding and most proteins fold spontaneously.

- Proteins have hydrophobic cores and hydrophilic surfaces
- Proteins often fold into domains
    - Domains often have discrete functions
- Protein structure is determined by its constitutive amino acid sequence
- Chaperones – proteins that assist the covalent folding or unfolding and the assemble or disassembly of other macromolecular structures

Protein Misfolding

Misfolded proteins can aggregate and form amyloid deposits; this can lead to diseases like Alzheimer's and Parkinson's.

- GroEL-GroES Complex
    - Prevent aggregation of misfolded proteins; gives them a chance to re-fold properly

Protein Processing

Proteins are often processed after translation to reach their final form.

- Some of the protein may be cleaved off
- Another chemical group might be added
    - Can active or inactive the protein
    - May further stabilize the protein
- May associate with a cofactor
    - Cofactor – non-protein compound that is required for a protein's function
- Proteins may have more than one stable conformation
    - Enzymes often change conformation during catalysis

Isoelectric Point (pI)

*The isoelectric point (pI) relates to the charge state of the **entire** molecule. The pI influences the separation of protein based on its charge.*

Protein Isolation

Protein isolation is sometimes needed to produce large quantities of purified proteins for subsequent use or to produce small quantity of protein for analytical purposes.

Chromatography

- Stationary phase is a porous matrix of beads (packed in a column)

- Mobile phase is buffer that percolates through the column by gravity

- Used for size exclusion

    o Larger molecules are excluded from the beads and move faster through the matrix

    o Smaller molecules move in and out of the beads which makes them slower

SDS-PAGE

- Sodium Dodecyl Sulfate (SDS)

    o A detergent to coat and denature the protein

    o Equalizes charge to mass ration

    o Equalizes shape – makes proteins linear

- Polyacrylamide Gel Electrophoresis (PAGE)

    o A gel matrix is used

    o Denatured protein moves toward the positive electrode when a current is applied

        ▪ Proteins separated based on size

        ▪ Larger molecules travel slower

Ion Exchange

- Separation of proteins based on charge

- Resin has a net charge

    o Molecules of opposite charge bind; others move through

- Salt is added to competitively displace proteins that become bound

CHAPTER 4: THE NON-CATALYTIC FUNCTIONS OF PROTEINS

Brief Overview

- Non-catalytic classes of proteins:
    - Binding proteins
        - Bind and release small molecules
    - Structural proteins
        - Help a cell maintain its shape, some help with movement
    - Motor proteins
        - Convert chemical energy into physical movement
- Main focus will be on myoglobin and hemoglobin because:
    - The characteristics of O_2 binding to myoglobin and hemoglobin are observed in many areas of biochemistry
    - Oxygen transport is essential in sustaining life

Myoglobin (Mb) and Hemoglobin (Hb)

Similarities between Mb and Hb

- Similar function (both bind O_2)
- Mb and Hb have similar tertiary structure
- Both Mb and Hb carry a heme prosthetic group
    - Prosthetic group - a nonprotein group that is forming part of a protein

Differences between Mb and Hb

- Mb is a monomer while Hb is a tetramer
- They are not very similar in primary structure
    - Only 18% identical
- O_2 binds differently to Hb and Mb

Binding of O₂ to Myoglobin

Myoglobin binds O₂ and delivers it to the mitochondria.

- K is a measure of the affinity of Mb for O₂

$$K_a = \frac{[MbO_2]}{[Mb][O_2]}$$

  - K decreases as binding affinity increases

- Y = fractional saturation (ratio of Mb bound to O₂: total Mb)

$$Y = \frac{[MbO_2]}{[Mb]+[MbO_2]}$$

- Hyperbolic plot

  - Binding increases rapidly until most of the molecules are saturated

  - Value of K is where Mb is half saturated (2.8 torr)

    - p50 is the designation for ½ saturation

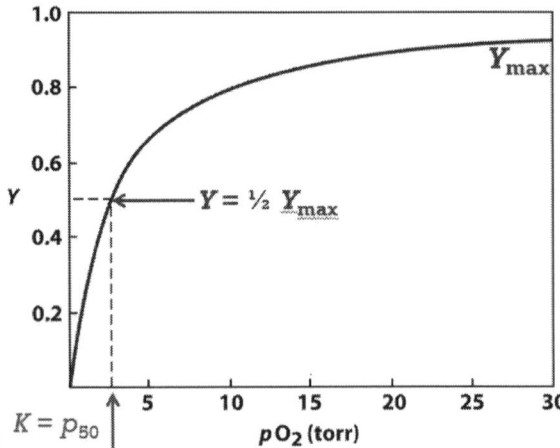

Plot of Myoglobin Binding to O₂:

Binding of O₂ to Hemoglobin

- Reaches the same saturation point (Ymax) as Mb

- Curve is sigmoidal – not hyperbolic

  - p50 values are very different (26 torr for Hb vs. 2.8 torr for Mb)

  - Hb affinity for O₂ is different at low vs. high pO₂

- At low pO_2, O_2 does not bind Hb easily

- As pO_2 increases, O_2 is more likely to bind to Hb

    - The 4th O_2 that binds, does so with 100 times the affinity as the 1st O_2

- This is an example of allostery

    - Binding at one site affects the affinity of other sites

- The difference in affinity is beneficial

    - Allows O_2 to bind in **lungs**, where pO_2 is high

    - Low affinity in **tissues**, allows for releases O_2 there

    - Transporting O_2 from lungs to tissues is the function of Hb

    - Mb affinity is high so that it can take up the released O_2

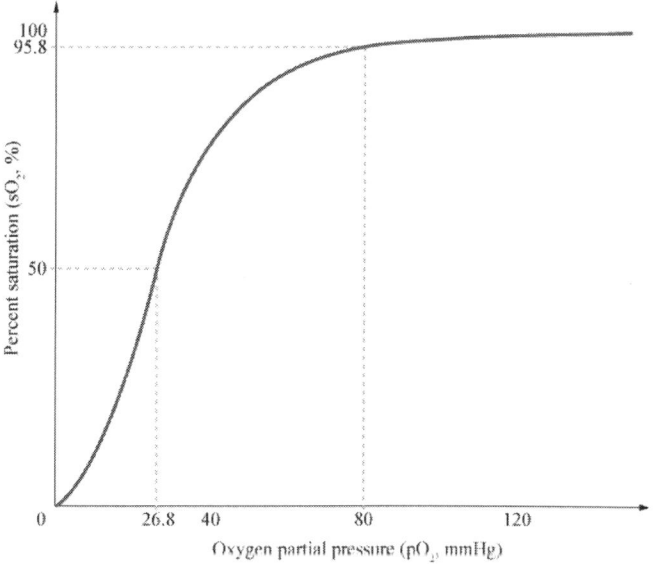

Plot of Hemoglobin Binding to O_2:

- Hemoglobin has 2 different states

    - One for low pO_2 and one for high pO_2

    - Tense (T) state: conformation not favorable for O_2 binding

    - Relaxed (R) state: conformation favorable for O_2 binding

Bohr Effect

A physiological phenomenon that states that hemoglobin's oxygen binding affinity is inversely related to acidity and the concentration of carbon dioxide.

- pH affects O_2 binding
    - $Hb \bullet H^+ + O_2 \leftrightarrow Hb \bullet O_2 + H^+$
    - As pH ↑ (i.e. [H^+]↓), O_2 binding favored (equilibrium shifts to the right)
    - As pH ↓ (i.e. [H^+]↑), O_2 release favored (equilibrium shifts to the left)

Regulating O_2 Binding to Hemoglobin

- 2,3-bisphosphoglycerate (BPG)
    - Binds to Hb in (T) state and stabilizes it
        - Binding cavity is too small in (R) state for BPG to bind
    - Favors O_2 release

Structural Proteins

Structural proteins are the most abundant class of proteins in nature.

- Microfilaments
    - Actin
- Intermediate filaments
    - Keratin
- Collagen
    - Microtubules
    - Tubulin

Actin Microfilament

There are two forms of actin filaments G-actin and F-actin.

- Globular protein with ATP binding site: G-actin
    - Monomeric

- Filamentous actin: F-actin
    - Double chain of subunits
    - Negative end has an ATP site
- Monomers are added together to form filament
    - Addition is more rapid at positive end
    - Driven by ATP hydrolysis
- Polymerization of actin is reversible
    - Can be used to extend and retract cellular processes
        - One method of cellular motility

Microtubules

Microtubules are used in the construction of cilia and flagella and are used to align and separate pairs of chromosomes during mitosis.

- Built from dimers of tubulin
- Forms protofilament (like actin)
- Forms a hollow cylinder, rather than a single filament
    - Makes it more stable

Formation of Microtubules

- α and β tubulin
    - Each monomer is capable of binding GTP nucleotide
    - GTP is blocked in α but solvent-exposed in β
    - Formation of microtubule causes hydrolysis of GTP in β but not in α
- 13 protofilaments line up side by side to form a tube
- Dimers can add at both ends
    - Positive side is assembled faster
    - Disassembly is also faster at the positive end

Intermediate Filaments

Intermediate filaments are exclusively structural proteins.

Keratin

- Basic unit is dimer of coiled α helices
- 7-residue repeat
  - #1 & #4 residues are nonpolar
- Two dimers associate to form tetramers, two tetramers to form octamer, etc.

Collagen

- Trimeric molecule
  - Holds cells together
  - Supports part of animal bones and tendons
- Every third residue is Gly
- 30% of rest is Pro and Hyp (hydroxyproline)

Motor Proteins

Myosin and Kinesin

- Basic structures are the same
- Their mode of action is different
- They have different functions in the cell

Myosin

Myosin is the thick filament and actin is the thin filament.

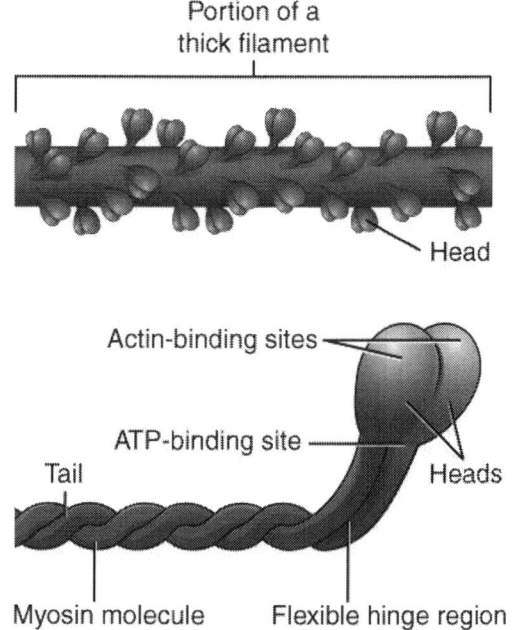

- Two polypeptide chains form the two heads and the long tail

- The heads have binding site for actin and for ATP

- Two light chains wound around each "neck" of the protein (4 total in dimer)

- The two heads act independently

- Myosin tails associate to form thick filament

    o With the heads sticking out

- Heads bind to actin filaments

Myosin Mechanism

- Reaction sequence begins with a myosin head bound to an actin subunit of the thin filament

    o ATP binding alters the configuration of the myosin head so that it releases actin

- Rapid hydrolysis of ATP to ADP + P_i triggers a conformational change that rotates the myosin lever and increases the affinity of myosin for actin

- Myosin binds to an actin subunit farther along the thin filament

- Binding to actin causes P$_i$ and then ADP to be released. As these reaction products exit, the myosin lever returns to its original position. This causes the thin filament to move relative to the thick filament (the power stroke)

  - ATP replaces the lost ADP to repeat the reaction cycle

  - Myosin heads do not remain bound to ATP/ADP and filament at the same time

- Power stroke is considered the end of the sequence

Kinesin

Motor proteins like kinesin move cargo between the cell body and the axon ends of neurons.

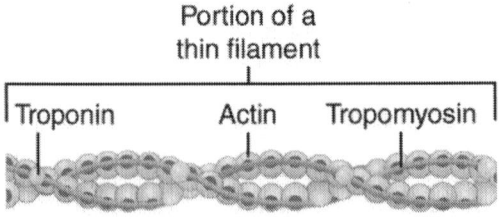

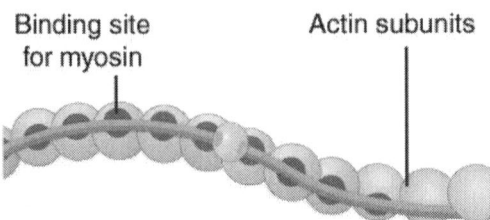

- Microtubule-associated motor protein

- Similar structure to myosin

  - Shorter neck then myosin

  - Light chains are located at the other end of polypeptides

- Two light chains bind some vesicle with the purpose of moving it along a microtubule

Kinesin Mechanism

- ATP binding to the leading head induces a conformational change in which the neck docks against the head. This movement swings the trailing head forward by 180° toward the positive end of the microtubule

    - This is the force-generating step

- The new leading head quickly binds to a tubulin subunit and releases its ADP. This step moves kinesin's cargo forward along the protofilament

- In the trailing head ATP is hydrolyzed to ADP + P_i. The P_i diffuses away, and the trailing head begins to detach from the microtubule.

- ATP binds to the leading head to repeat the reaction cycle

CHAPTER 5: ENZYMES

The Need for Biological Catalysts

Some chemical reactions need to happen more readily than other, and this can be accomplished in a few ways.

- Raise the temperature
    - But living organisms have a tolerable range and this is not always possible
- Increase the [reactants]
    - But the reactant may be scarce and it may be difficult to get more
- **You can also add a catalyst**
    - Speeds up a reaction but is not consumed in process
    - Makes them very effective and useful

Enzymes as Catalysts

- Each enzyme has an active site where catalysis takes place
    - Active site - a region on the enzyme which binds substrate
        - Usually a pocket or groove located on the surface of the enzyme
        - The pocket or groove determines the specificity of the enzyme

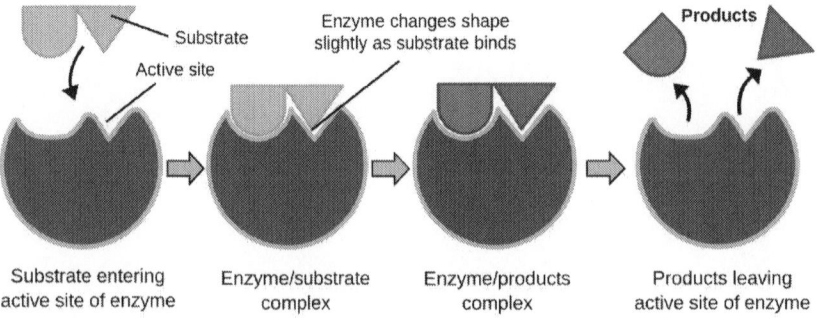

| Substrate entering active site of enzyme | Enzyme/substrate complex | Enzyme/products complex | Products leaving active site of enzyme |

Classes of Enzymes

- Oxidoreductases – involved in oxidation-reduction reactions
- Transferases – involved in transfer of functional groups
- Hydrolases – involved in hydrolysis reactions

- Lyases – perform group elimination to form double bonds

- Ligases – involved in bond formation coupled with ATP hydrolysis

- Isomerases – involved in isomerization reactions

Naming enzymes

- Enzymes are often named for the substrate they act on

    o Name often tells you what it does

- Name most often ends in "-ase"

- May have more than one version, multiple enzymes may be catalyzing the same reaction

    o Isozymes operate in different tissues or expressed under different conditions but catalyze the same chemical reaction

    o Often differ in catalytic properties

<u>Chemistry of Catalysis</u>

Catalysts do not change the direction of a chemical reaction and they have no effect on equilibrium!

- Function by lowering the activation energy, which speeds up the reaction

- As the reaction progresses; reactants become products

    o Depicted as a reaction coordinate diagram

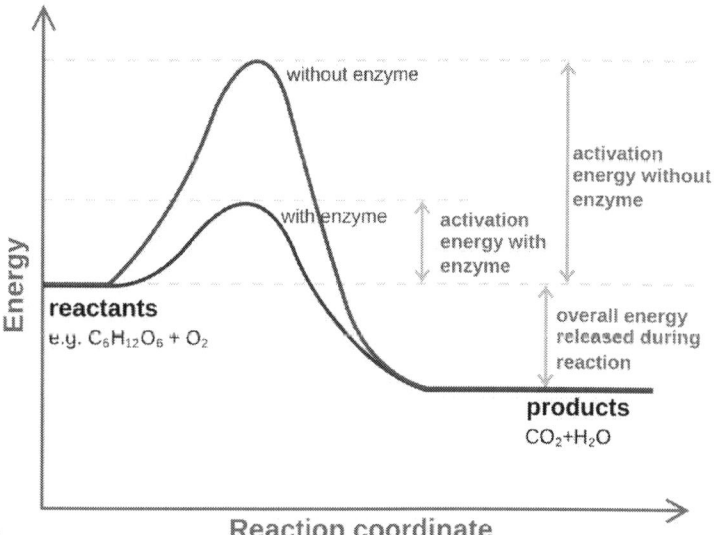

Reaction Coordinate Diagram:

- o Progress of the reaction is indicated on the x-axis

- o Free energy (G) is indicated on the y-axis

- Reactants pass through the transition state (‡) and become products

  - o Enzymes increase the reaction rate by binding tightly to the transition state and stabilizing it

Spontaneous vs. Non-spontaneous Reactions

- Spontaneous if ΔG_{rxn} negative

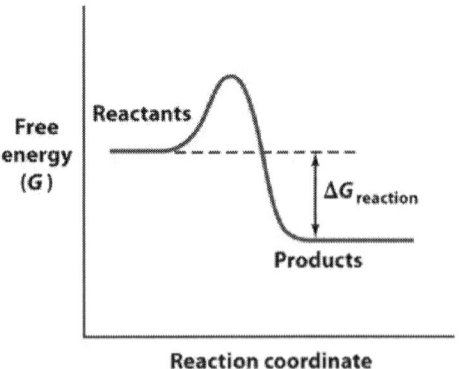

General Spontaneous Reaction Diagram:

- Non-spontaneous if ΔG_{rxn} positive

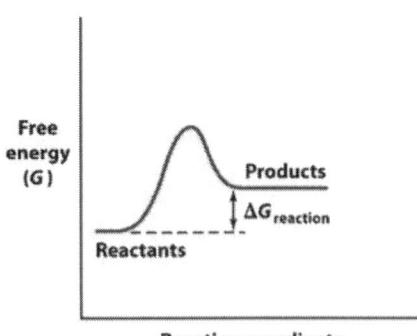

General Non-spontaneous Reaction Diagram:

- Enzymes (catalysts) lower the energy barrier

  - o Make it easier to reach the transition state

Acid-Base Catalysis

- Involves transfer of H^+

- Some use acid, some base, some both

Covalent Catalysis

- Enzyme cycles between 2 states
  - Covalent bond forms between enzyme and substrate in the 1st step
  - Bond is broken in the 2nd step to regenerate the original enzyme
- 2-part reaction – 2 energy hills
  - One hill to form Schiff base ($X_1^{\ddagger}$)
  - One hill to decompose it ($X_2^{\ddagger}$)
  - Intermediate still less stable (higher G) than reactants or products

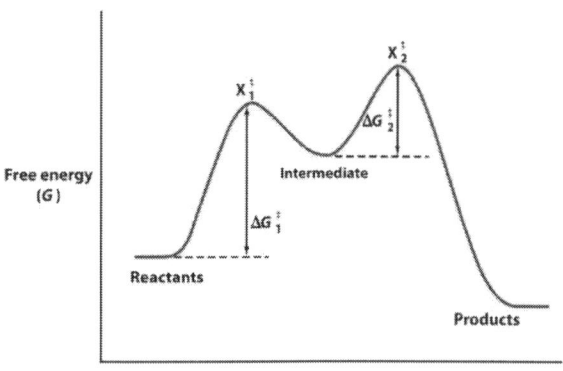

Covalent Catalysis Reaction Diagram:

Metal Ion Catalysis

- Mediate oxidation-reduction reactions
  - Movement of 1 or 2 electrons
- May interact directly with the substrate
- May stabilize the charge of the transition state

Substrate Specificity

- Enzymes vary in their specificity pocket
  - Vary in the amino acids that comprise the pocket
    - Controls the chemical environment
    - Also controls the dimensions of the pocket

- Examples:
  - **Chymotrypsin**
    - Gly residues makes for a deep pocket (remember: glycine is the simplest amino acid with a single H atom for its side chain)
    - Pocket can fit an aromatic ring
  - **Trypsin**
    - Pocket has the same dimensions as chymotrypsin, but different side chain at bottom
      - Has Asp which is negatively charged instead of Gly
    - Asp (-) readily binds Lys (+) or Arg (+)
  - **Elastase**
    - Bulkier groups on the walls of the pocket
      - Create a smaller pocket
    - Binds small nonpolar side chains

Other Enzyme Effects

- Proximity effects - get reacting groups close enough
- Orientation effects - get reacting groups in proper orientation

Activation of Chymotrypsin

- Proteases are often synthesized as inactive precursors
  - Activated when and where they are needed
  - Some activate themselves (autoactivation)
    - Active form of the enzyme activates the inactive form
- Trypsin converts chymotrypsin from inactive to active form
  - Chymotrypsin then further activates itself
  - Benefit of this is that there is no time delay to synthesize the protein

Biochemical Cascades

Cascades are common in biological systems (e.g. clotting cascade in blood) and they are useful in producing big effects in a short amount of time.

CHAPTER 6: ENZYME KINETICS AND INHIBITION

***Note*:** *This chapter contains many brackets for example "[substrate]" which is shorthand for "concentration of substrate." In general terms, "[x]" is meant to be read as "concentration of x."*

Velocity of a Reaction

- As a reaction progresses, [Substrates] ↓ and [Products] ↑ linearly

  o Rate at which this happens is the velocity of the reaction

$$v = -\frac{d[S]}{dt} = \frac{d[P]}{dt}$$

- More of the enzyme present, the faster the rate

Velocity vs. [Substrate]

- At constant [Enzyme], velocity changes in nonlinear fashion as [Substrate] ↑

- Hyperbolic plot

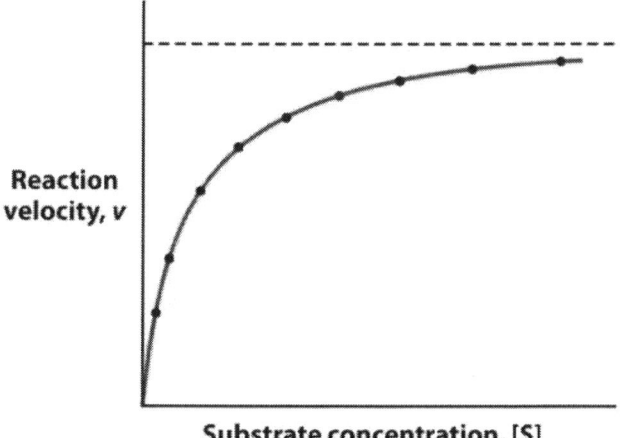

- At low [Substrate], velocity ↑ linearly (steep slope)

- At high [Substrate], activity levels off (shallow slope)

  o Indicates enzyme saturated with substrate

  o Indicates formation of enzyme-substrate (ES) complex

$$E + S \leftrightarrow ES \rightarrow E + P$$

- At max velocity, all enzymes are working at max speed

Rate Equations

- Consider the unimolecular reaction: A → B

$$v = -\frac{d[A]}{dt} = k[A]$$

- Rate is directly proportional to the amount of the reactant

- Rate constant (*k*) is first-order

  - Only depends on one reactant (substrate)

- Bimolecular reaction: A + B → C

$$v = -\frac{d[A]}{dt} = -\frac{d[B]}{dt} = k[A][B]$$

  - Rate constant is second-order

    - Depends on concentration of two reactants

  - Case 1: two reactants make a product

  - Case 2: two molecules of one reactant make a product

Michaelis-Menten Kinetics

Model of enzyme kinetics, describes the rate of enzymatic reactions.

$$v = \frac{d[P]}{dt} = \frac{V_{max}[S]}{K_m + [S]}$$

- V_{max} = K_2[Enzyme]

  - Maximum rate achieved by the system

  - Point at which enzymes are fully satuarated with substrate

  - Maximum rate of product formation

- $K_{max} = (k_{-1} + k_2)/k_1$
    - K_1 = rate of enzyme-substrate complex formation
    - k_{-1} = rate of enzyme-substrate complex dissociation
    - k_2 = rate of product formation
    - Substrate concentration when the reaction rate is half of V_{max}
    - Low value means high affinity of an enzyme to its substrate
- $K_{cat} = V_{max}/[Enzyme]$
    - This is the turnover number - number of substrate molecules that are converted to product per enzyme per unit time

Steady State Kinetics

State in which the concentration of an enzyme substrate complex very quickly reaches a constant value; enzyme substrate complex is formed as rapidly as another enzyme substrate complex disappears.

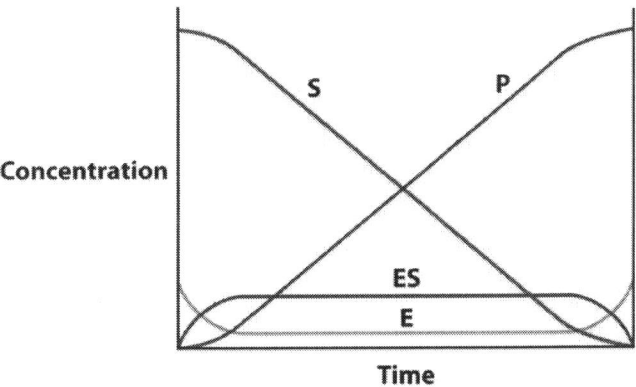

- [Substrate] declines at approximately a constant rate over time
- [Product] increases at approximately a constant rate over time
- [Enzyme-Substrate Complex] increases initially and then doesn't change over time
- Assumption is that the substrate is in much higher concentration than the enzyme

Catalytic Efficiency

These are factors that influence catalytic efficiency:

- Rate of electronic rearrangements needed to make the product

- How often the enzyme collides successfully with the substrate

  - Also known as the diffusion-controlled limit

- Catalytic perfection

  - Enzyme catalyzes reaction as rapidly as it encounters a substrate

Lineweaver-Burk Plot

- Linear plot

- Slope = K_m/V_{max}

- Y-intercept = $1/V_{max}$

- X-intercept = $-1/K_m$

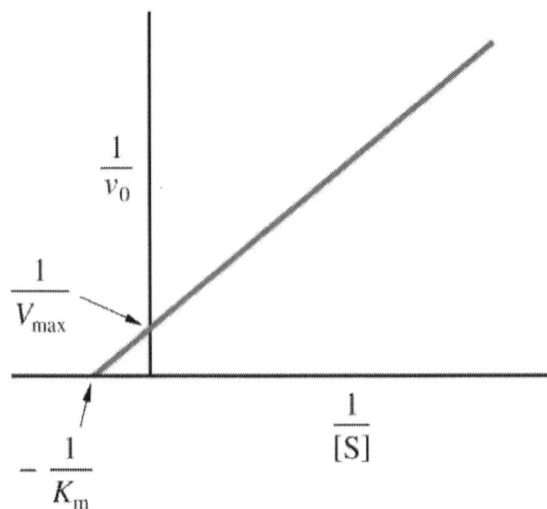

Lineweaver-Burk equation:

$$\frac{1}{v_0} = \left(\frac{K_m}{V_{max}}\right)\frac{1}{[S]} + \frac{1}{V_{max}}$$

Non-Michaelis-Menten Enzymes and Reactions

There are some reactions and enzymes that don't fit into the simple Michaelis-Menten kinetics.

- Cellular reactions that involve 2 substrates

- Bi-Bi reactions – 2 substrates and 2 products

  - Ordered Sequential

    - Substrate bind in specific order

- - Both substrates bind, then products are released
  - Products are released in a specific order
  - Random Sequential
    - Binding of substrates and release of products is not ordered
  - Ping-pong
    - 1st substrate binds; 1st product is formed and released
    - 2nd substrate binds to form 2nd product and regenerate the enzyme
- Multi-step reactions
  - K_{cat} becomes very complicated in reactions with multiple steps
- Allosteric Enzymes
  - Cooperativity - presence in one active site affects catalytic activity at other active sites
  - Usually a multi-subunit complex

Irreversible Enzyme Inhibition

- Covalent group attachment
  - Attach covalently, deactivate the enzyme, may be removed by enzymatic activity to reinstate the function of the enzyme
- Binding of suicide substrate
  - Begins to react but can't complete the reaction
  - Bind like normal substrate
  - Get stuck in the active site

Types of Reversible Enzyme Inhibition

Competitive Inhibition

- Compound resembles substrate or transition state
  - Those that resemble the transition state are usually more potent
- Occupies the active site
- Adding more substrate can out-compete inhibitor

- No change in catalysis (once a substrate binds you get product at the same rate)
  - No change in V_{max}
  - Increase K_m
  - K_{cat} is the same
- Substrate concentration curve
  - X-intercept closer to 0
  - Steeper slope

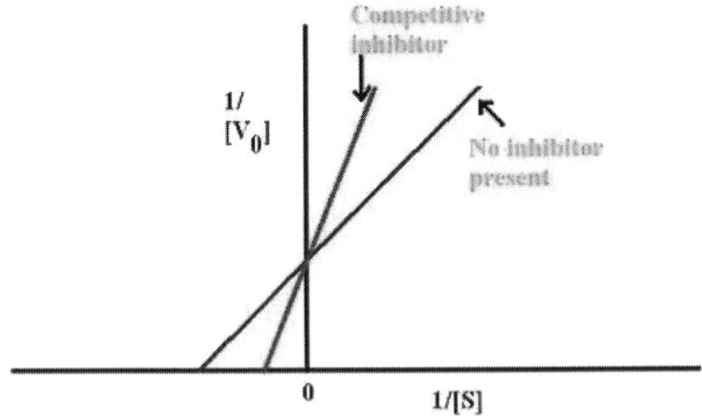

Noncompetitive inhibition

- Binds to both the enzyme and enzyme substrate complex
- No effect on affinity, no change in K_m
- V_{max} is lower

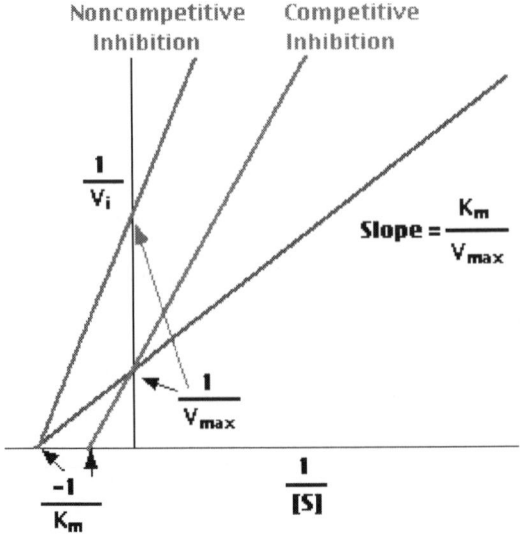

Mixed inhibition

- K_{cat} decreases
- K_M increases
  - It may decrease, but it's rare
- V_{max} is lower

Uncompetitive inhibition

- Inhibitor binds only to enzyme substrate complex (ES)
- Binding of substrate may open up binding site for inhibitor
  - Inhibitor may bind adjacent to substrate or only in its presence
- Binding of inhibitor lowers the [ES], favoring the formation of the ES complex
    - Lowers K_M
    - Lowers V_{max}
    - Prevents catalysis, so lowers k_{cat}
  - Substrate concentration curve
    - X-intercept farther from 0
    - Y-intercept increased

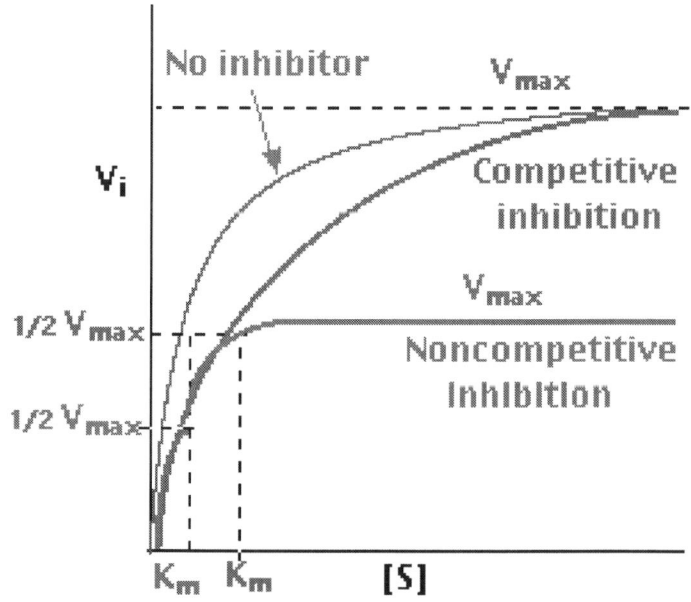

Allosteric Regulation

Allosteric regulation is the regulation of a protein by binding of an effector molecule.

- Ligand binds to site other than active site

    o Ligand – substance that forms a complex with a biomolecule to serve a biological purpose

- Binding can inhibit or activate the enzyme

- Produces sigmoidal plots

Regulating Enzyme Activity

These are possible ways of regulating enzyme activity:

- Change in subcellular location

    o Bring enzyme to substrate or sequester it away

- Enzyme can be activated by ionic signal

    o Change in pH or release of stored Ca^{2+}

- Covalent modification

    o Add a group to activate or deactivate

    o Remove a group to reverse the effect

Chapter 7: Lipids and Membranes

Lipids (Fats) Overview

- Mostly hydrophobic molecules
- Differ from amino acids and nucleotides in that:
  - They cannot form polymers
  - Do not have defining functional groups
- Some, but not all are amphipathic
  - Contain both polar and nonpolar regions
- Functions
  - Cellular barrier
    - Control movement into and out of cell
  - Control of membrane fluidity
  - Place for proteins to dock and travel
  - Waterproofing agents (wax)
  - Signaling agents

Types of Lipids

Fatty Acids

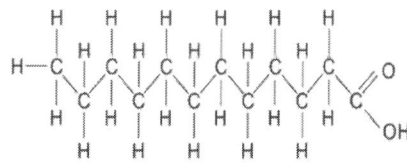

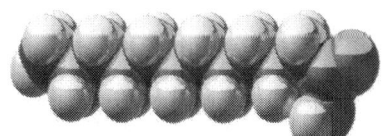

Fatty Acid Example: Lauric acid ($C_{12}H_{24}O_2$)

- Long chain of carboxylic acids
- Even-numbered chains are most common

- Saturated fatty acids

    o Tails saturated with H atoms

    o No double bonds

- Unsaturated fatty acids

    o Contain at least one double bond

    o Double bonds usually have *cis* configuration

- Degree of saturation determines how they "pack"

    o Double bonds produce kinks in the chain

        ▪ Makes it harder for molecules to stack

- Fatty acids are not typically found in free form

    o Usually esterified to glycerol

Triacylglycerols (Triglycerides)

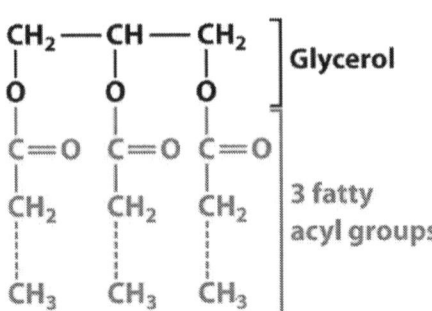

General Structure of Triacylglycerols:

- Glycerol backbone

- Esterified to 3 fatty acids

    o Fatty acid chains may be different

        ▪ Provides variety of structure/function

- Do not form bilayers (not a part of biological membranes)

- Aggregate into globules (stored for energy)

Glycerophospholipids

- Glycerol backbone

- 2 fatty acids + phosphate-derivative head group

- o They are amphipathic because of the head group
- Abundant in membranes
- Phospholipases
  - o Hydrolyze phospholipids
  - o May release phospho head group or fatty acid chain

Sphingolipids

- Sphingosine backbone
- One acyl chain already present as part of backbone
- Second fatty acid chain linked to sphingosine via amide bond
- May contain a phosphate-head group
- Head group could also have saccharide head groups
  - o Cerebroside – monosaccharide head group
  - o Ganglioside – oligosaccharide head group
- Sphingomyelin - part of myelin sheath around nerves

Sterols

- Constructed from 5-Carbon units like isoprene
- Rigid structure

Lipid Bilayer

- Forms spontaneously due to the hydrophobic effect
- Formed from glycerophospholipids and sphingolipids
  - o Has to do with the geometry of those molecules
  - o Cholesterol can partition in membrane
    - ▪ Cannot form bilayer by itself
- Fatty acids and triglycerides cannot form bilayers
  - o Their geometry is wrong

- 25-30 Å thick
  - Thickness varies depending on length of acyl chains and how they interact
    - Also depends on the size of head groups

Membrane Fluidity

The bilayer is very fluid, it has no defined geometry.

- Fatty acid mobility
  - Higher at high temperatures
  - Smaller chain length = more fluidity
  - Saturation = more fluidity
  - Melting Temperature
    - Longer chains = higher melting point
    - Saturation = lower melting point
      - Saturation has greater effect than chain length
- Cholesterol influences fluidity
  - Restricts lipid movement and decreases fluidity in fluid state
  - Prevents close packing of lipids in crystalline form
  - Works like a fluid buffer
    - Helps lipids resist melting at high temperature
    - Helps lipids resist crystallization at low temperature
- There may be local regions of membrane that are nearly crystalline called lipid rafts
  - Contain cholesterol and sphingolipids
  - More rigid microdomains within membrane

Lipid Bilayer Translocation

- Membrane proteins can move laterally easily
- Move between monolayers less frequently

- May require a protein for translocation (translocases)
- Interleaflet movement is essential for asymmetry
    - The two leaflets have different lipid compositions

Membrane Proteins

- Membrane is ~50% protein
    - Depends on cell type
- Types of membrane proteins:
    - Integral membrane proteins
        - Span bilayer
        - Require detergent to separate them from the membrane
    - Anchored proteins
        - Carry lipid anchor that integrates in membranes
    - Peripheral protein
        - Associates with integral proteins or lipid head groups

Integral Membrane Proteins

- Transmembrane helices
    - May be single-pass (1 helix) or multi-pass (more than 1 helix)
    - Composed of hydrophobic residues
    - Need 20 amino acids to span the bilayer
    - Polar aromatic residues often occur where helix approaches polar head groups

- Transmembrane β sheets
  - Must have several β strands to satisfy hydrogen bonding of backbone
  - Form a β barrel

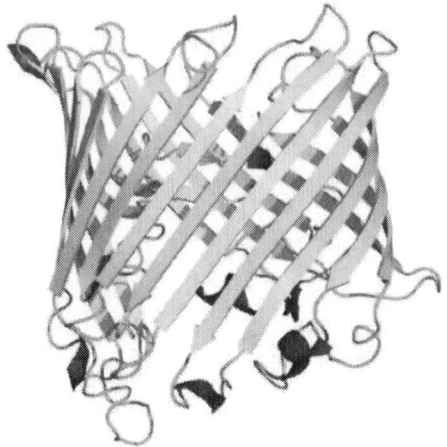

    - Smallest have 8 strands
      - Narrow cavity is more limiting (selective)
  - Exterior surface has hydrophobic residues

Anchored Proteins (Lipid-linked Proteins)

- Soluble proteins
  - Except the lipid anchor
- Some may also have transmembrane regions
- May be anchored inside or out

<u>Fluid Mosaic Model</u>

Model usually represented as membrane proteins that float randomly in a sea of lipids. This is mostly true, but membrane proteins are not always entirely free to move.

- Movement is restricted by:
  - Interactions with other proteins
  - Interactions with cytoskeletal proteins

- Membrane protein may:
  - Be immobile
    - Attached to cytoskeleton
  - Mobile within a small area
  - Freely mobile
- Proteins as well as lipids may be asymmetrical in membrane

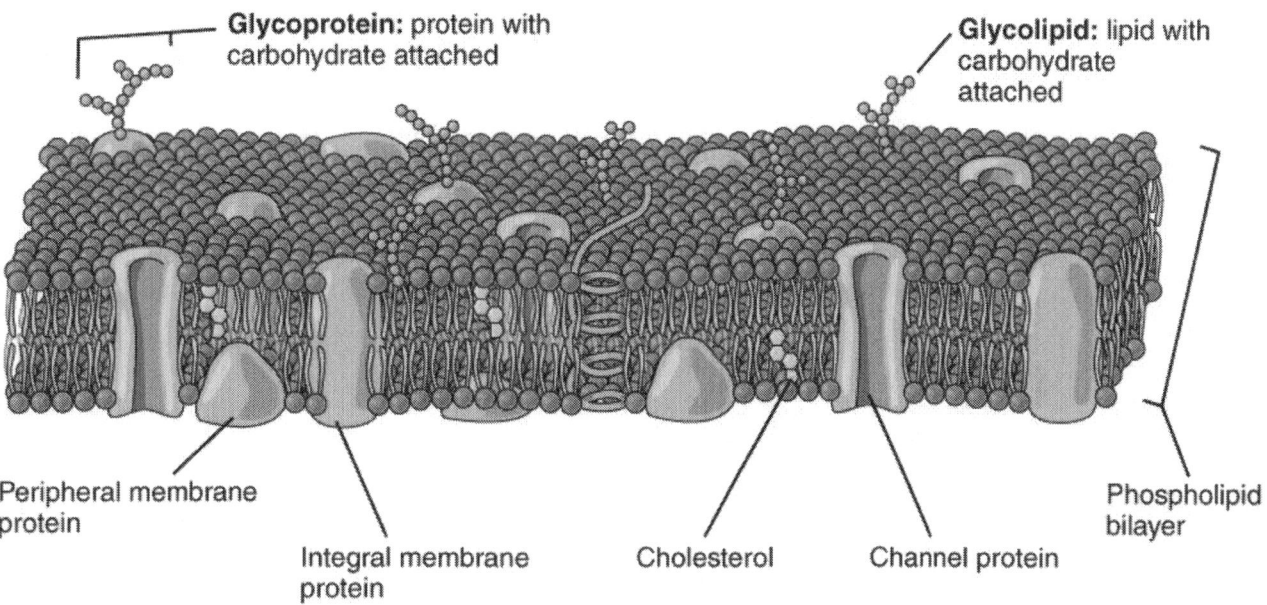

Glycoprotein: protein with carbohydrate attached

Glycolipid: lipid with carbohydrate attached

Peripheral membrane protein

Integral membrane protein

Cholesterol

Channel protein

Phospholipid bilayer

CHAPTER 8: MEMBRANE TRANSPORT

Principles of Membrane Transport

- Movement of ions results in membrane potential
    - Concentration gradient across membrane
    - $[Na^+]_{in} < [Na^+]_{out}$
    - $[K^+]_{in} > [K^+]_{out}$
- Ions move spontaneously ($-\Delta G$) down their concentration gradient

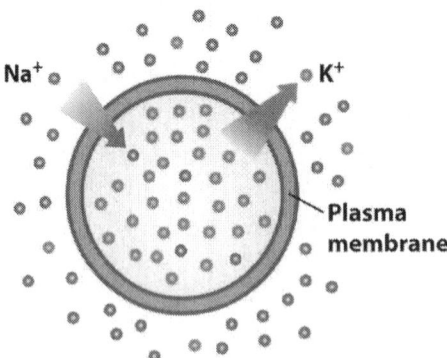

Active vs. Passive Transport

- Active transport - energy is needed (nonspontaneous process) for a transport protein to pump a substance against its concentration gradient
- Passive transport - diffusion across a membrane, no energy needed (spontaneous)
    - Flow down concentration gradient
        - Concentration gradient – gradual difference of concentration over a distance in a particular direction
    - Passive diffusion (simple diffusion)
        - Diffusion – net movement of a substance down a concentration gradient
        - No pore or transport protein is needed

- o Facilitated diffusion

  - ▪ Transport proteins are used to pass material into or out of the cell

  - ▪ Transport proteins in facilitated diffusion do not use energy

    - • Substance being transported travels down its concentration gradient

Passive Transport

Passive transport occurs when ions and molecules move down their concentration gradient.

$$\Delta G = RT \ \ln \frac{[X]_{in}}{[X]_{out}}$$

- • ΔG negative (spontaneous) only if [X]out > [X]in

- • When membrane potential is present:

$$\Delta G = RT \ \ln \frac{[X]_{in}}{[X]_{out}} + ZF\Delta\Psi$$

  - o R = Gas constant (8.3145 J/K*mol)

  - o T = Temperature in K (C + 273)

  - o Z = Net Charge Per Ion

  - o F = Faraday Constant (96.485 J/V*mol)

  - o Monovalent Cation at 20 C, ΔΨ = 0.058 log ([ion]in/[ion]out]

Passive Transport Porins

Porins are β barrels and all known porins are timers.

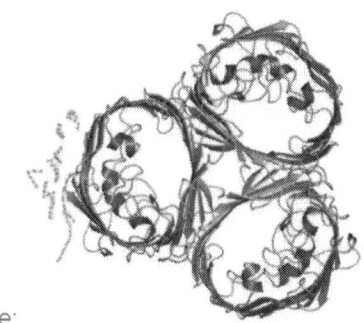

Porin Example:

- Each barrel functions as a transporter that is always open

- Can transport in either direction

- Hydrophilic side chains line the inside of barrels

Ion Channel Selectivity

Ion channels are more complex in structure than porins and this allows them to be more selective with what they transport.

- Multimers of α-helical segments

- Ion passageway in central axis where helices meet

- K^+ channel 10,000x more likely to transport K^+ than Na^+

    - Due to geometry of "selectivity filter" (Na^+ is smaller)

- 4 backbones fold so carbonyls (–C=O) project into the core

    - K^+ is a good fit but Na^+ is too small to coordinate with the carbonyls

    - Tyr residues prevent contraction of the pore so it doesn't accommodate the smaller Na^+ cation

Aquaporins

Aquaporins are very common in nature and they allow for rapid transport of water.

AQP1 family

- Homotetramer with carbohydrate chains on extracellular surface

- Each monomer is mostly 6 α-helices that span bilayer + 2 shorter helices that are within bilayer

- Each subunit is a water pore
  - ~3 Å (water = 2.8 Å)
  - Lined with hydrophobic residues except for 2 Asn
    - Prevents H^+ from getting in as H_3O^+
      - This would cause a change in the pH
  - Asn residues hydrogen bond with water to transport a single molecule
    - Rather than a network of hydrogen bonded molecules

Gated Channels

Gated channels are not always open, they first have to receive and then respond to a signal. A signal can be a change in pH or the binding of a ligand or small molecule.

- Neuronal K^+ channel is voltage-gated
  - Responds to depolarization
  - Helices near intracellular side move to expose pore on the extracellular side
  - After channel first opens, N-terminal segment of the protein moves to block pore
    - This is why it doesn't immediately reopen
- Mechanosensitive channels
  - Open in response to membrane tension
  - Set of α helices move past each other to alter packing arrangement
  - A small pore even when open
    - Opening lined with bulky hydrophobic residues
  - Control osmotic balance in bacteria
  - Thought to be sensors for touch, hearing, balance

Glucose Transporter

Transport through the glucose transporter is accomplished by eliciting a conformational change and it works by a "rocker" mechanism.

- Mechanism:

    - Glucose binds on one side

    - Elicits conformational change

    - This exposes molecule to the other side

    - Example of facilitated (passive) transport

- Transport proteins are similar to enzymes in that:

    - They can become saturated with substrate

    - They are subject to competitive and other types of inhibition

Types of Transporters

Transporters may work through passive or active transport and are classified by the way in which they function.

- Uniporter - Moves one substance at a time

- Cotransport – a single ATP-powered pump that actively transports one solute and indirectly drives the transport of other solutes against their concentration gradients

    - Symporter - Moves two substances in the same direction at the same time

    - Antiporter - Moves two substances in opposite directions at the same time

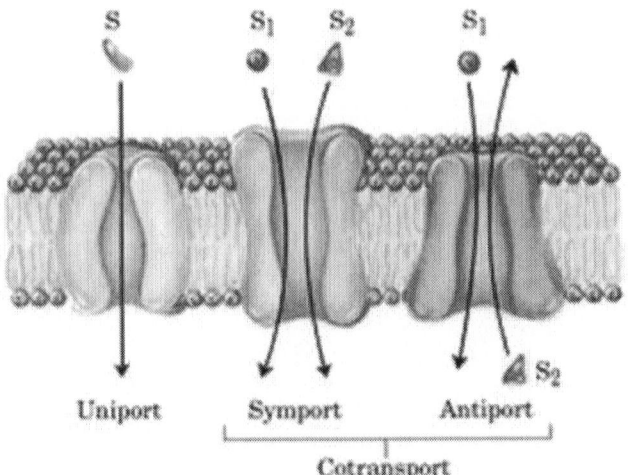

Types of Transporters:

Na, K-ATPase Reaction Cycle

This is not the same Na⁺ and K⁺ transporters directly involved in axon potential. Net effect of the reaction cycles is that 1 ATP is used pump 3 Na⁺ ions out and bring 2 K⁺ ions in (K⁺ IONS ARE LARGER).

Reaction Cycle:

- 3 intracellular Na^+ ions bind

- ATP binds (Na^+ binding triggers ATP binding)

- A phosphoryl group is transferred from ATP to an Asp side chain of the pump

    o ADP is released (phosphorylation triggers conformational change)

- The protein conformation changes, exposing the Na^+ binding sites to the cell exterior

    o Na^+ ions dissociate

- Two extracellular K^+ ions bind

    o Binding stimulates dephosphorylation

- The aspartyl phosphate group is hydrolyzed

    o P_i is released

- The protein conformation changes, exposing the K^+ binding sites to the cell interior

    o K^+ ions dissociate

Active Transporters

Na⁺/K⁺-ATPase

- Large α subunit that has 10 transmembrane helices

- Cytoplasmic domain carries aspartate residue

- Membrane region carries binding site for **K⁺** ions

ABC Transporters

- <u>A</u>TP-<u>B</u>inding <u>C</u>assette

- Use ATP energy to pass material into cell

    o Transports material against its gradient

- Solute binding protein binds solute; complex binds transporter
- Binding domain is stimulated to bind and hydrolyze ATP
  - Phosphate not transferred to enzyme, as in Na^+/K^+-ATPase
- This opens channel for solute to pass through

Secondary Active Transport

*A form of active transport that couples the movement of an ion **down** its concentration gradient with the movement of another ion **against** its concentration gradient.*

- Symporter can move glucose into a cell with Na^+
- Na^+ gradient already established by Na^+/K^+-ATPase
- Glucose can then exit the cell as internal [glucose] increases

Acetylcholine Release

Acetylcholine (a neurotransmitter) is stored in synaptic vesicles.

Mechanism:

- When an action potential reaches the axon terminus it causes voltage-gated Ca^{2+} channels to open
- Increase in intracellular Ca^{2+} ion concentration triggers the fusion of synaptic vesicles with the plasma membrane
  - Acetylcholine is released into the synaptic cleft
- Acetylcholine binding to receptors on the surface of the muscle cell leads to a muscle construction
  - Signal is short-lived because any acetylcholine remaining in the synaptic cleft is rapidly degraded

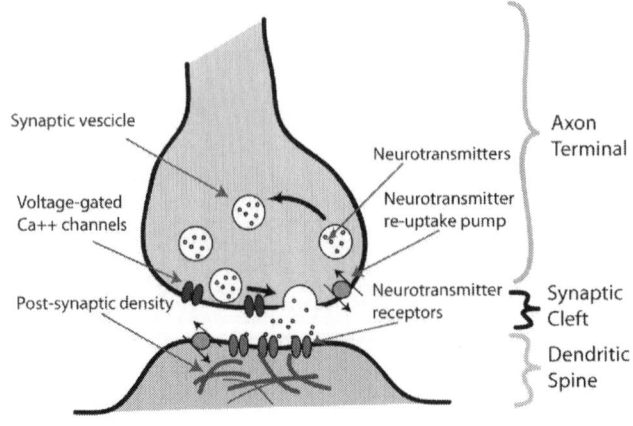

Membrane Fusion

Proteins called "SNAREs" participate in tethering two membranes.

- Soluble N-ethylmaleimide-sensitive-factor attachment-protein receptor
    - Integral membrane proteins
    - 2 SNAREs from the plasma membrane + 1 SNARE from the vesicle form a complex
- Mutual contact is required
    - Ensures that the correct membranes fuse
- Spontaneous process
- Membrane fusion is important in many cellular processes
    - Processing proteins, etc.

CHAPTER 9: SIGNALING

Signal Transduction

- Requires a receptor to receive a signal and a ligand to provide it
- Ligand interacts with receptor to generate response
    - Always involves a conformational change in receptor
- Many types of ligands
    - Peptides
    - Lipids
    - Amino acid derivatives
    - Hormones

Ligand Binding

Ligand is any substance that forms a complex with a biomolecule to serve a biological purpose.

- Ligands bind with high affinity to receptors
- Strength of binding: R + L $\longleftrightarrow$ R•L

$$K_d = \frac{[R][L]}{[R \cdot L]}$$

- K_d is ligand concentration at which receptors are half-saturated with ligand (½ of the receptors are bound)
- Agonist - ligand that binds to receptor and elicits a biological effect
- Antagonist - binds to receptor but does not trigger a response

G-Protein-Coupled Receptors (GPCRs)

- GPCRs are 7-transmembrane receptors
    - Many are lipid-linked
    - Ligand binding site comprised of helical core and loops

- Binding of ligand induces conformational change in transmembrane protein (1st messenger)
    - Helices shift at the ligand binding site
- Allows it to interact with G protein (inactive form)
    - Releasing GDP and binding GTP activates the G protein
- G protein can activate other proteins/enzymes
    - These enzymes often produce small molecules – **2nd messengers**
        - These are diffusible messengers
            - Initiate changes in cellular activities by diffusing through the cell

Receptor Tyrosine Kinases

- Binding of ligand activates the receptor
- Transfers phosphoryl group to tyrosine residue on itself (autophosphorylation)
    - May activate or inhibit the protein or enzyme
- May initiate a cascade of kinase-activation events
- Results in changes in metabolic activity or gene expression

Regulation of PKA

PKA is a family of enzymes whose activity is dependent on cellular levels of cyclic AMP (cAMP).

Protein Kinase A

- 2 regulatory (R) subunits + 2 catalytic (C) subunits
    - R subunits block substrate binding sites in C subunits
- Binding of cAMP to R domains releases C domains

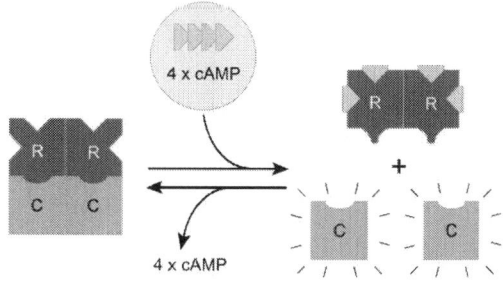

- - Allows them to bind substrate
- Phosphorylation of PKA - Activation loop of PKA
  - Carries Thr residue that is phosphorylated in active PKA
  - Phospho-Thr (-) interacts with Arg side chain (+)
  - Positions Asp in active site
    - Interacts with substrate and positions it for phosphoryl transfer from ATP
  - Loop blocks active site if not phosphorylated

Desensitization

Desensitization allows cells to adapt and respond to future changes (allows for return to resting state). It is a multi-step process that produces an amplified response (a small signal can have a big effect).

- GPCR may be desensitized by phosphorylation and subsequent binding of arrestin
  - Prevents interaction with G protein

Second Messengers

- G protein activates phospholipase C
  - Cleaves phosphatidylinositol bisphosphate (PIP_2)
  - Forms inositol triphosphate (IP_3) + diacylglycerol (DAG)
    - IP_3 very polar → can diffuse through the cytoplasm
    - DAG very nonpolar → diffuses in membrane
- IP_3 and DAG are both second messengers

PIP_2 Pathway

- IP_3 triggers opening of Ca^{2+} channels
  - Initiates kinase cascade
- DAG diffuses in membrane to bind protein kinase C
  - This docks PKC at the membrane and activates it
    - PKC phosphorylates Ser/Thr residues

- o Starts another kinase cascade

- o PKC also requires Ca^{2+} for full activation

Receptor Tyrosine Kinases

Most receptors are monomers that dimerize upon ligand binding.

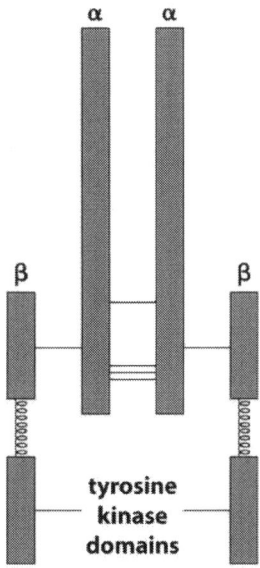

Insulin Receptor Example

- 4 polypeptide segments ($\alpha_2\beta_2$), held together by sulfate bonds

    - o β subunits contain tyrosine kinase domains

- Binding of one insulin molecule pulls 2 α subunits together

    - o Only 1 insulin bound per receptor

- Ligand binding induces conformational change

    - o 2 Tyr-kinase domains brought together

    - o Phosphorylate each other

    - o Removes activation loop from active site allowing substrate to bind

Lipid Hormone Signaling

Some hormones in the body are lipids.

- Nonpolar messengers with the ability to diffuse through membrane barrier

- Bind to receptors inside the cell instead of cell-surface receptors

- Bind to Hormone Response Elements (HREs) in DNA

    - 2 identical DNA sequences, separated by a few base pairs

- Function as transcription factors to control gene expression

CHAPTER 10: CARBOHYDRATES

Carbohydrate Nomenclature

- General formula: $(CH_2O)_n$; $n \geq 3$

- Also known as saccharides

    - Monosaccharides – one unit

    - Di-, tri-, etc.: ≥ 2 units

    - Larger polymers: polysaccharides

    - Can be modified with other groups to contain N, P, etc.

Classifying Carbohydrates

- Classified by functional group carrying carbonyl

    - Aldose = aldehyde

    - Ketose = ketone

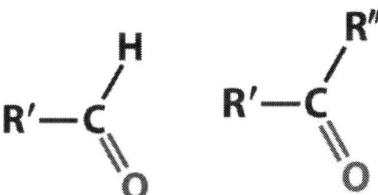

Aldehyde   Ketone

- Classified by the number of carbon atoms

    - Triose (3)

    - Tetrose (4)

    - Pentose (5)

    - Hexose (6), etc.

- The anomeric carbon is always bound to 2 O atoms

- Pyranose - a 6-membered ring

- Furanose – a 5-membered ring

- Most carbs are chiral

  - 4 different groups on at least 1 carbon

  - Nearly all monosaccharides have number of stereoisomers

    - Stereoisomers – isomeric molecules that have the same molecular formula and sequence of bonded atoms but that differ in their three-dimensional orientation

  - Named by comparison of penultimate carbon to glyceraldehyde

    - Look at the last chiral carbon

    - Compare to L- and D-glyceraldehyde

      - These are enantiomers (mirror images)

  - Fischer projections make it easy to spot enantiomers

    - Enantiomers have identical physical properties

    - Can only distinguish them by

      - Interactions with other chiral substances

      - Rotation of plane-polarized light

    - Most naturally-occurring sugars are D enantiomers

- Epimers - carbs that differ in configuration around only 1 carbon other than the anomeric one)

  - Diastereomers

  - Non-mirror images

Sugar Representation

The three common representations are Fischer Projection, Haworth Projection, and Chair Conformation (pictures are in that order).

α Anomer   Glucose

Reducing vs. Nonreducing Sugars

Reducing Sugars

- Reducing sugars are carbohydrates that can reduce oxidizing agents
  - Carry reactive carbonyl group
- Sugars which form open chain structures with free carbonyl group
- Can reduce metal ions
  - They themselves become oxidized

Nonreducing Sugars

- Carbohydrate that is not oxidized by a weak oxidizing agent
  - Usually very stable in water
- Common non-reducing sugar is sucrose

Polysaccharides

Polysaccharides are chains or polymers of carbohydrates.

- Formed by glycosidic bonds
  - Aka glycan chains
- Can be used to modify other molecules (e.g. glycoproteins or glycolipids)
- Polysaccharides are common fuel-storage molecules
- Can be joined in a variety of ways
  - Multiple –OH groups with different bonding arrangements makes this possible

Common Disaccharide (Lactose)

Disaccharides occur in nature most commonly as breakdown products of polysaccharides

Lactose

- Secreted in milk of lactating mammals

- Galactose-β-1,4-glucose

- Joined by glycosidic bond between #1 C of galactose and the #4 C of glucose (left to right)

- Different C's joined to form the bond would result in different disaccharides

Fuel-storage Carbohydrates

Starch

- Polymers of glucose joined by α(1-4) linkages

Amylose

- Linear (unbranched) form of starch

- Linkages create "kink" in structure

- Leads to helical shape

Amylopectin

- Same linkages as starch but have shorter chains

- Includes α(1-6) linkages every 20-30 residues to generate branched polymer in starch

- Larger molecule

Glycogen

- Polymer that resembles amylopectin, but branches are about every 12 residues

- Highly branched form allows it to be rapidly assembled or disassembled

Structural Carbohydrates

Cellulose

- Polymers of glucose joined by β(1-4) linkages

    o Remember that starch has α(1-4)

- This difference is responsible for creating large differences in structure

    o Instead of forming compact granules, it forms extended fibers

    o Provides rigidity and strength for plant cell walls

        ▪ Extensive hydrogen bonding network within and between adjacent chains

Chitin

- β(1 → 4) linkages of glucose derivative

    o *N*-acetylglucosamine

- Part of the exoskeletons of insects and crustaceans

- Part of the cell walls of many fungi

Biofilms

- Extracellular matrix of carbohydrates

- Attaches to surfaces and harbors community of microbes

- Microbes produce and maintain the biofilm

    o Combination of glucuronate and *N*-acetylglucosamine

Glycoproteins

Glycoproteins are proteins that carry carb groups.

- Eukaryotic glycoproteins are usually *N*-linked to Asn or *O*-linked to Ser or Thr

- *N*-glycosylation

    o Chain of 14 residues that is added as a polypeptide comes off ribosome in rough endoplasmic reticulum

    o At Golgi apparatus, glycosidases remove some residues and glycosyltransferases add new monosaccharides

        ▪ Gives tremendous variety and specificity to molecules

- *O*-glycosylation
    - Tend to be built one residue at a time in Golgi apparatus
    - Do not undergo processing
    - More extensive network than *N*-linked glycoproteins
- Purpose of glycosylation
    - Highly hydrophilic and conformationally flexible
    - Occupy large volume above protein's surface
    - May be protective or may help stabilize a protein
    - May guide molecular chaperones in proper folding of proteins

Proteoglycans

Proteoglycan are glycoproteins that are mostly carbs.

- Carry attachment sites for large *O*-linked glycosaminoglycans
    - Attached to protein portion at Ser or Thr
- Most often repeating disaccharide of amino sugar (N-acetyl) and uronic acid (sugar with COO^-)
- Sulfate group may be added post-synthesis
- Protein portion may be transmembrane or lipid-linked
- Sugar chains usually on extracellular side
    - Many hydrophilic groups – highly hydrated
- Bacterial cell walls are made of peptidoglycan
    - Determines overall shape
    - Provides structural support to prevent cell rupture (molecular girdle)

CHAPTER 11: METABOLISM AND BIOENERGETICS

Catabolism, Anabolism, and Metabolism

Energy for growth comes from the chemical reactions of the cell.

- Catabolism

    o Breaking down larger molecules to release free energy and small molecules

- Anabolism

    o Using energy to build cell components

- Metabolism

    o Balance between catabolism and anabolism

Products of Digestion

Human digestion reduces biomolecules to monomers.

- Starch is digested by amylases

    o Amylases are found in salivary glands

- Proteins are digested by proteases

    o Proteases are secreted in the stomach and pancreas

- Fatty acids are digested by lipases

    o Lipases are made in the pancreas and secreted in the small intestines

Processing Nutrients

- Fatty acids

    o Reform into triglycerides inside the cell

    o Some fatty acids are linked to cholesterol

- o Chylomicrons form and are released into bloodstream via lymphatic system
    - Chylomicrons are lipoproteins containing cholesterol esters, triacylglycerols, and proteins
    - Deliver lipids to different cell types
    - Mostly processed by liver
- Water-soluble compounds (amino acids, sugars) leave intestinal cells and enter the portal vein leading to liver
- Liver receives the bulk of nutrients
    - o Catabolizes them, stores them, or releases them back into bloodstream
    - o Liver is the main processing plant in the body

Storing Nutrients

Storing Fatty Acids

- Fatty acids are stored in the form of triacylglycerols (large globules) in adipocytes

Storing Carbohydrates

- Almost all cells can metabolize monosaccharides
- Liver and muscle store glucose as glycogen
    - o Branching of glycogen means rapid storage and rapid release
- Glucose can be catabolized to 2-C acetyl units and converted to fatty acids to be stored as triacylglycerols

Storing Amino Acids

- Amino acids are not intended for long-term storage
- Proteins can be catabolized to supply energy
    - o **Excess** free amino acids following digestion may be catabolized to provide energy
    - o Amino acids that are part of cellular protein are not catabolized except during starvation
- Excess amino acids can be broken down and converted to carbohydrate and stored as glycogen or to acetyl units and stored as fat

Mobilizing Stored Fuels

Energy reserves are deployed when dietary intake is insufficient.

- Polymers are converted to monomeric units

- Liver breaks down glycogen to release glucose

    - Catalyzed by glycogen phosphorylase

    - Residues released from end by phosphorolysis

    - Phosphate removed before it released into the blood

- When glucose is low, adipose tissue mobilizes fat stores

- Lipase releases fatty acids from triacylglycerols

    - Not water-soluble

        - Bind to circulating proteins

    - Last energy reserve to be tapped (normal conditions)

        - Higher potential energy source than carbohydrates or amino acids

        - Since it is the most "valuable reserve" with regards to energy potential, your body only taps into it when it is necessary

Protein Degradation

"Stored" amino acids are not utilized except during fast, they require more processing than carbohydrates. Two ways to degrade proteins:

Lysosome

- Organelle that contains proteases and hydrolytic enzymes (low pH)

- Proteins enclosed in membranous vesicle

    - Fuse with lysosome

- Used to degrade particles from within (autophagy) or outside (phagocytosis or endocytosis)

- No waste is released into the cytoplasm

Proteasome

- Barrel-shaped, multiprotein complex in the cytoplasm

- Protein to be degraded is 1st tagged with ubiquitin

    o Need 4 or more to tag for degradation

        ▪ Makes sure only the right protein is degraded

    o Protein is unfolded as it enters the proteasome

        ▪ Protein unfolded through ATP hydrolysis

    o Attached to Lys side chain of protein

Metabolic Pathways

Only a few steps are needed to convert polymers to monomers but many steps are required to catabolize monomers or anabolize them.

- Metabolic pathway - series of chemical reactions needed to break down or build up monomers

    o Each step generally involves a different enzyme

    o Always regulated in some way

- The purpose of many steps

    o Convert large energy source into multiple smaller sources or energy carriers

        ▪ Less chance that energy is wasted that way

        ▪ Smaller "packets" means cells don't spend more than what they need

        ▪ Multiple steps means multiple intermediates or metabolites

Redox Reactions

- Recall definitions of oxidation and reduction

    o Oxidation - loss of electrons

    o Reduction - gain of electrons

- Catabolism of amino acids, monosaccharides, and fatty acids involves oxidizing carbon

    o Replace C—H with C—O

- C—H share electrons equally; O atom in C—O pulls electrons (e-) from C

- Anabolism of amino acids, monosaccharides, and fatty acids involves reducing carbon

- Fatty acids have many methylene carbons that undergo oxidation

- Carbohydrates have (CH_2O) carbons that undergo oxidation

Redox Cofactors

- Electrons can get passed from metabolites to enzyme cofactors such as NAD^+ or $NADP^+$

- Ubiquinone (Q) Carrier

  - Can accept 1 or 2 electrons, in sequential fashion

- Cofactors are "recycled" through oxidative phosphorylation

Complexity of Metabolism

Metabolic pathways are often connected; substrates in one pathway are products of another.

- Pathway activity is highly regulated

  - Cell regulates flow of metabolites

  - Shut down some pathways; turn on others

- Not every cell uses every pathway

Free Energy and Metabolic Reactions

Energy is the capacity to do work, in biological systems; it is the capacity for chemical change.

- For reaction: A + B $\leftrightarrow$ C + D

$$K_{eq} = \frac{[C]_{eq}[D]_{eq}}{[A]_{eq}[B]_{eq}}$$

- At equilibrium, there is no net change in concentration

  - Doesn't mean that [A] = [B] or that [A][B] = [C][D]

- When not at equilibrium, reactants move to reach equilibrium values

- Standard free energy change for reaction, $\Delta G^{\circ\prime}$

$$\Delta G^{\circ\prime} = -RT \ln K_{eq}$$

  - o R = gas constant (8.3145 J/K/mol); T = temp (Kelvin)
- Standard (°) according to chemistry
  - o 25°C (298 K)
  - o 1 atm pressure
  - o 1 M
  - o This is not sufficient for biological systems
- Prime (') specifies the reaction under standard biochemical conditions
- In living cells, conditions are rarely "standard"
- Define actual change in free energy, ΔG as:

$$\Delta G = \Delta G^{\circ\prime} + RT \ln \frac{[C][D]}{[A][B]}$$

- Equation above shows that the spontaneity of the reaction (ΔG) is related to actual concentrations, not just $\Delta G^{\circ\prime}$
  - o Reaction with $+\Delta G^{\circ\prime}$ can still occur, depending on reactant concentration
- Thermodynamically unfavorable reactions can proceed if they are coupled with a favorable reaction
  - o Costs energy to produce Glc-6-P
  - o Couple with ATP hydrolysis
  - o Net effect is thermodynamically favorable ($-\Delta G$)
  - o This is a very common theme in biochemical systems

CHAPTER 12: GLUCOSE METABOLISM

Pathways of Glucose Metabolism

- Glycolysis
    - "Lysis" indicates that this process is a breakdown of glucose
    - Glucose is broken down into the end product pyruvate
- Gluconeogenesis
    - Results in the generation of glucose
    - Used when the supply of glycogen is exhausted
- Glycogen synthesis and degradation
    - Important in storing glucose for the long-term and recovering it later
- Pentose phosphate pathway
    - Used to generate pentose (5-C sugar)
    - Starts with glucose

Metabolic Pathways

Different metabolic pathways have many properties in common.

- Each step is a different chemical reaction
    - Also catalyzed by a different enzyme
- Free energy released or consumed is transferred to "carriers" like ATP and NADH
- Rate of pathway flux is controlled by changing activity of individual enzymes
- Multiple steps means greater energy recovery
    - In one step, most of the energy generated would be lost as heat
- Two general purposes of catabolic pathways
    - Convert energy from a molecule into a more usable form
    - Create intermediates to be used in other pathways

Free Energy Changes in Glycolysis

- Three steps have large negative values of ΔG (Steps 1, 3, and 10)

  - *Potential* flux-control points

  - Slower steps

    - Pathway can only go as fast as its slowest step

- Other steps have $\Delta G \approx 0$

  - Near-equilibrium reactions

  - Can accommodate flux in either direction

  - Direction controlled by concentrations of substrates & products (mass action ratio)

The 10 Steps of Glycolysis

- Net equation of glycolysis:

  - Glucose + 2 NAD^+ + 2 ADP + 2 P_i → 2 pyruvate + 2 NADH + 2 ATP

- 10-step oxidation pathway

  - Steps 1-5: Energy investment phase

  - Steps 6-10: Energy payoff phase

Step 1

- Hexokinase adds a phosphate group to glucose

- Step uses 1 ATP molecule

- $\Delta G^{\circ\prime}$ = (-) 16.7 kJ/mol

  - Highly negative ΔG value – this step is irreversible

    - Commits the metabolite to pathway

    - Blocks glucose from transport out of cell

Step 2

- Phosphoglucoisomerase converts glucose-6-phosphate into its isomer fructose-6-phosphate

- $\Delta G°' = + 2.2$ kJ/mol but $\Delta G = - 1.4$ kJ/mol

  - This is due to reactant concentrations

  - Low ΔG value means the step is reversible

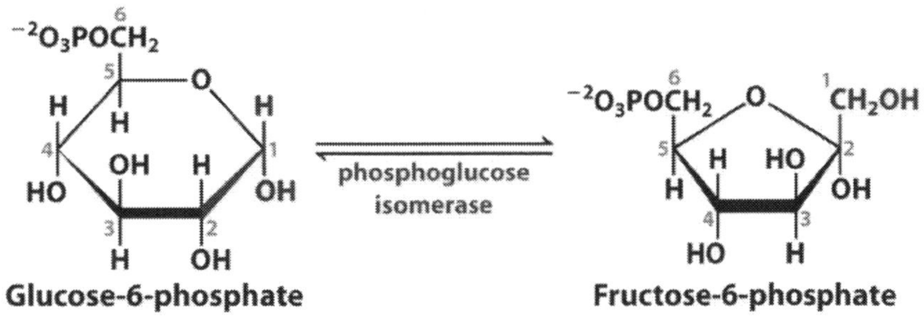

Glucose-6-phosphate ⇌ phosphoglucose isomerase ⇌ Fructose-6-phosphate

Step 3

- Phosphofructokinase (PFK) transfers a phosphate group to fructose-6-phosphate to form fructose 1,6-bisphosphate

- Step uses 1 molecule of ATP

- $\Delta G°' = -17.2$ kJ/mol

  - Highly negative ΔG value – this step is irreversible

- This is the flux-control point (rate determining step) of the pathway

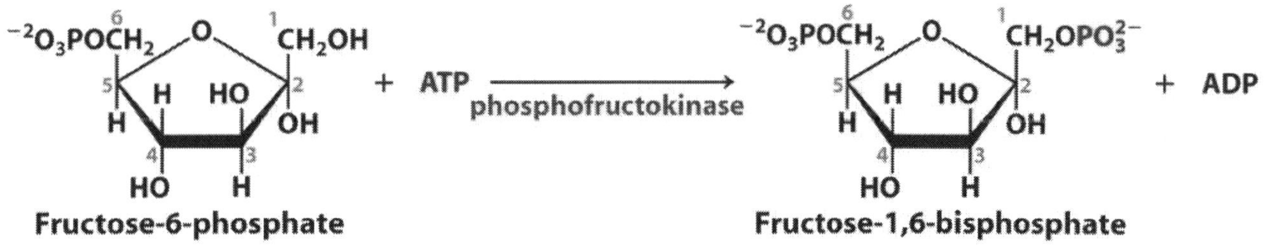

Fructose-6-phosphate + ATP → phosphofructokinase → Fructose-1,6-bisphosphate + ADP

Step 4

- Aldolase splits fructose 1,6-bisphosphate into dihydroxyacetone phosphate (DHAP) and glyceraldehyde-3-phosphate (Gly-3-P)

- $\Delta G°'$ = +22.8 kJ/mol

  - But ΔG < 0, due to rapid consumption of both products

    - Mass action ratio pulls the reaction forward

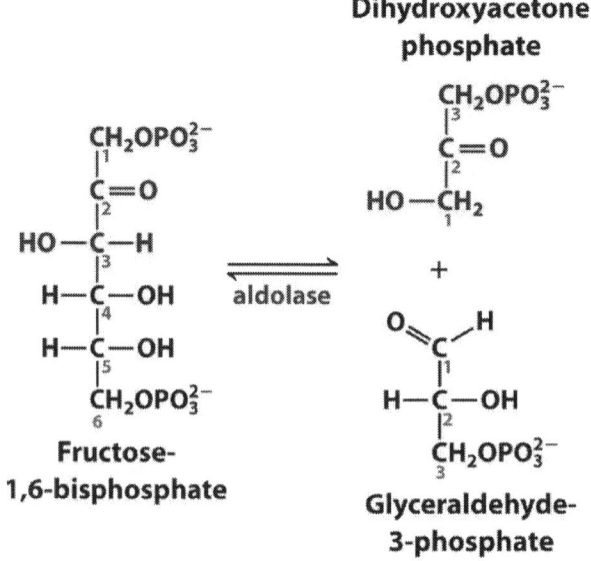

Step 5

- Triose phosphate isomerase rapidly converts DHAP to Gly-3-P

- ΔG = +4.4 kJ/mol

  - Consumption of Gly-3-P drives it forward

- This step is needed because it is simpler to send two identical molecules through the remainder of pathway because it simplifies regulation

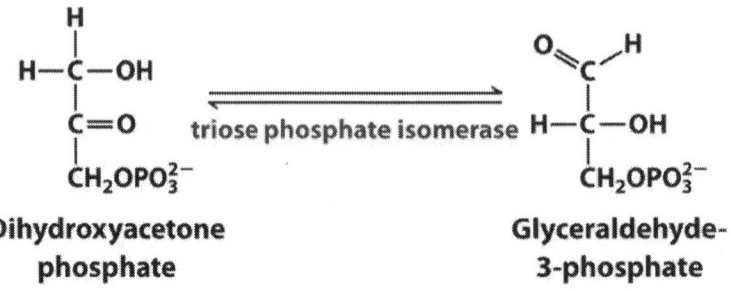

Step 6

Note*: This step and the steps that follow happen for both molecules of Gly-3-P

- Triose phosphate dehydrogenase transfers hydrogen from Gly-3-P to NAD^+ to form NADH

    o Enzyme then adds a phosphate group to the oxidized Gly-3-P

- Phosphate comes from P_i (HPO_4^{2-})

    o No ATP is consumed in this step

- NADH produced

- $\Delta G^{\circ\prime}$ = +6.7 kJ/mol

    o Pulled forward by next reaction

    o Energy recovered in the form of NADH

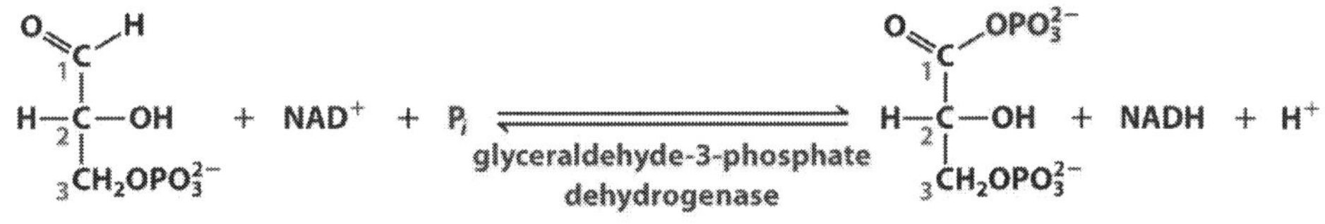

Glyceraldehyde-3-phosphate + NAD^+ + P_i ⇌ (glyceraldehyde-3-phosphate dehydrogenase) 1,3-Bisphosphoglycerate + NADH + H^+

Step 7

- Phosphoglycerokinase transfers a phosphate from the product of the previous step to ADP to make ATP

- $\Delta G^{\circ\prime}$ = -18.8 kJ/mol

    o This step helps pull Step 6 forward which has $\Delta G^{\circ\prime}$ value of +6.7 kJ/mol

    o This is an example of coupling a favorable with an unfavorable reaction

- Recover some of the energy used in the form of ATP

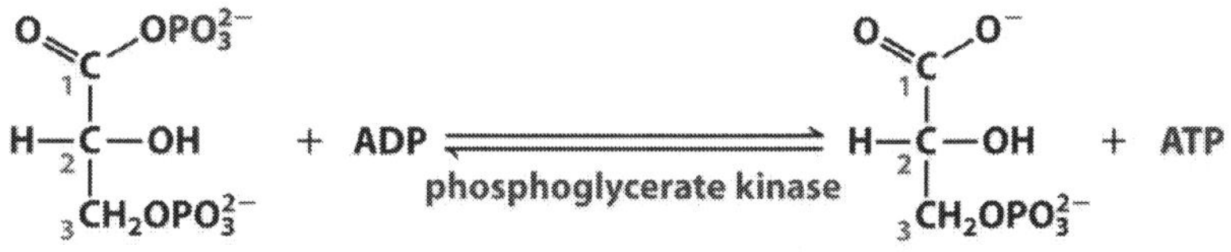

1,3-Bisphosphoglycerate + ADP ⇌ (phosphoglycerate kinase) 3-Phosphoglycerate + ATP

Step 8

- Phosphoglyceromutase relocates the phosphate from 3-phosphoglycerate from C_3 to C_2 to form 2-phosphoglycerate

- Phosphoryl needs to be on C_2 for what happens in Step 9

3-Phosphoglycerate   2-Phosphoglycerate

Step 9

- Enolase removes H_2O from 2-phosphoglycerate to form phosphoenolpyruvate (PEP)

2-Phosphoglycerate   Phosphoenolpyruvate

Step 10

- Pyruvate kinase transfers a phosphate from PEP to ADP to make ATP

- There are 2 parts to the reaction
    - 1st part: hydrolysis of phosphoryl is $\Delta G^{\circ\prime}$ = -16 kJ/mol
        - But for ADP → ATP, $\Delta G^{\circ\prime}$ = +30.5 kJ/mol

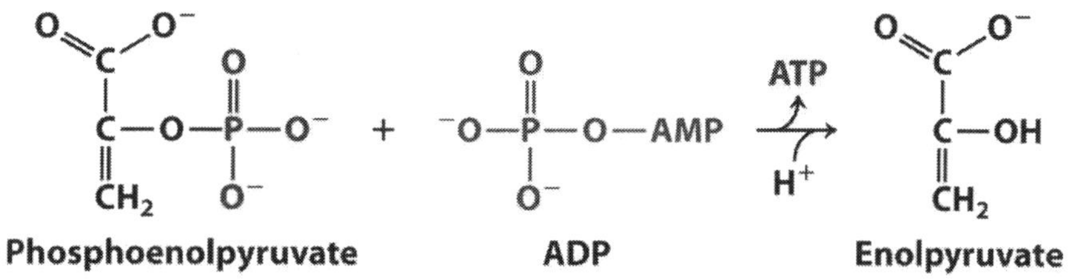

Phosphoenolpyruvate   ADP   Enolpyruvate

- 2nd part: tautomerization (shift of H atom) is $\Delta G°' = -46$ kJ/mol
  - This drives synthesis of ATP ($\Delta G°' = -31.5$ kJ/mol)
    - Recover energy (x2)
- Phase 1 cost: -2ATP; Phase 2 gain: +4 ATP
  - Net: 2 ATP gained

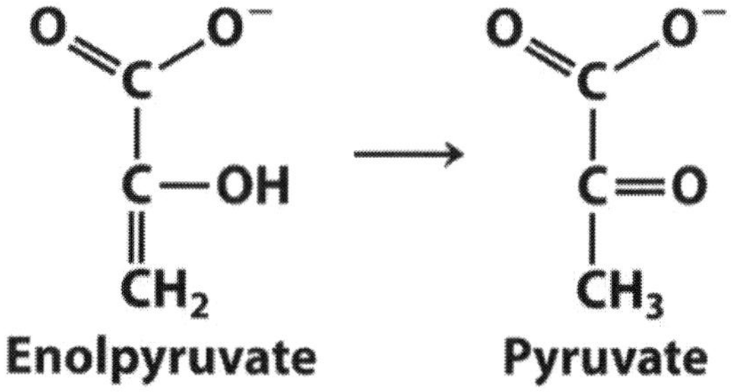

Enolpyruvate   Pyruvate

Regulation of PFK

PFK (phosphofructokinase) is the enzyme that carries out the reaction in step 3 of glycolysis. Step 3 is the rate determining step (the flux-control point) so in order to regulate glycolysis you have to regulate PFK.

- Recall: Three steps have large negative values of ΔG
  - Steps 1, 3, and 10
  - These 3 steps are irreversible
- Why is PFK (step 3) the actual control point?
  - Glucose can also enter the pathway as Glc-6-P
    - Bypasses 1st step – so Step 1 is not a good control point
  - Step 10 is irreversible
    - But it's the last step – not a good control point

In bacteria:

- PFK is regulated by allosteric effectors (+ and -)
- ADP is an activator
    - As [ADP] increases, more ATP is needed to turn on the pathway
- PEP (phosphoenolpyruvate) is an inhibitor
    - As [PEP] increases, indicates that too much product is present
        - Shuts down pathway

In mammals:

- Fructose-2,6-bisphosphate (Fru-2,6-P_2) is an activator: turns on the pathway
    - When blood glucose high, insulin is produced
        - Tells the cells there is a need to process glucose
    - Insulin stimulates PFK-2 to make Fru-2,6-P_2
    - Fru-2,6-P_2 activates PFK-1
    - More glucose sent through glycolytic pathway

Gluconeogensis

Gluconeogenesis can be thought of as the reversal of glycolysis because we start with 2 molecules of pyruvate and end with glucose.

- Anabolic pathway
    - Consumes energy
    - Doesn't operate in all cells
    - Liver uses this to make glucose for other cells
- Reversal of glycolysis
    - Conversion of 2 pyruvates to 1 glucose
    - Some steps in glycolysis we can reverse ($\Delta G \approx 0$)
    - Some steps we can't (large negative ΔG)
        - 4 different enzymes are needed for these steps
- 1st two steps of gluconeogenesis reverse reaction of pyruvate kinase (Step 10 of glycolysis)
    - Pyruvate carboxylated from bicarbonate (use energy from ATP)

- o Then decarboxylated and phosphorylated to give PEP (uses GTP)
- Use glycolytic enzymes to get back to Frc-1,6-P$_2$
- Cannot reverse the reaction catalyzed by PFK – need a different enzyme
  - o Fructose bisphosphatase hydrolyzes phosphoryl group
  - o $\Delta G°' = -8.6$ kJ/mol
- Phosphoglucoisomerase used next
- Cannot reverse reaction of hexokinase – need a different enzyme
  - o Glc-6-phosphatase removes phosphoryl group to form glucose

Need for Regulation

If glycolysis and gluconeogenesis occurred simultaneously, there would be a net consumption of ATP.

glycolysis	glucose + 2 ADP + 2 P$_i$ → 2 pyruvate + 2 ATP
gluconeogenesis	2 pyruvate + 6 ATP → glucose + 6 ADP + 6 P$_i$
net	4 ATP → 4 ADP + 4 P$_i$

- Glycolysis and gluconeogenesis are regulated based on the cell's needs
- Control point is fructose bisphosphatase
  - o Recall that Frc-2,6-P$_2$ is allosteric activator of PFK
  - o It is an inhibitor of fructose bisphosphatase
  - o As [Frc-2,6-P$_2$] decreases, PFK not activated and fructose bisphosphate not inhibited
  - o A single compound can control flux through two opposing pathways in a reciprocal manner

Storing Glucose

Glucose is stored in the liver and other tissues as glycogen.

- Glc-6-P converted to Glc-1-P
- Glc-1-P activated by adding UMP from UTP
  - o Reversible phosphoanhydride exchange ($\Delta G \approx 0$)
  - o PP$_i$ (inorganic pyrophosphate) rapidly hydrolyzed

- $\Delta G^{\circ\prime}$ = -33.5 kJ/mol
- Makes formation of UDP-glucose irreversible
- Glycogen synthase transfers glucose to end of glycogen branch with release of UDP
- Transglycosylase branching enzyme creates branch point

Releasing Stored Glucose

Glycogenolysis is the breakdown (lysis) of glycogen.

- Phosphorolysis yields Glc-1-P
- Mutase converts it to Glc-6-P
- Phosphatase removes phosphate
- Glucose transported out of (liver) cell into bloodstream, for transport to other tissues
- In other tissues, Glc-6-P enters glycolysis at Step 2
  - No need to add phosphate to glucose – saves 1 ATP

Pentose Phosphate Pathway

- Purpose of pathway
  - Converts Glc-6-P to **Rib-5-P**
  - Oxidative pathway in all cells
    - All cells need to make DNA and RNA
  - Generates **NADPH** instead of NADH
  - Anabolic
- Net reaction:

$$glucose\text{-}6\text{-}phosphate + 2\,NADP^+ + H_2O \rightarrow$$
$$ribose\text{-}5\text{-}phosphate + 2\,NADPH + CO_2 + 2\,H^+$$

Step 1

- Hydride (H^-) transferred to $NADP^+$

- Anomeric C oxidized

- This is an irreversible transfer

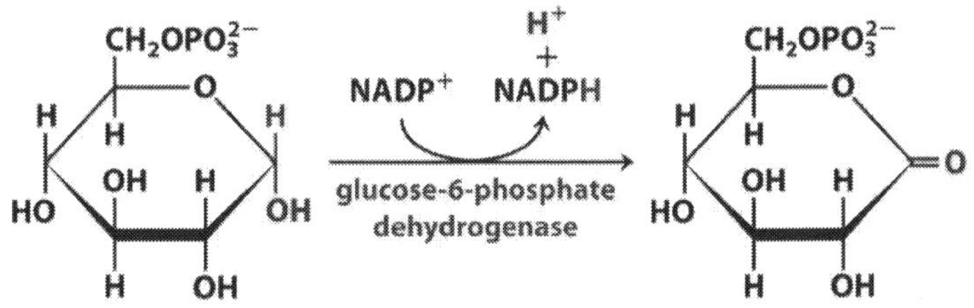

Glucose-6-phosphate → 6-Phosphoglucono-δ-lactone

glucose-6-phosphate dehydrogenase

Step 2

- Lactone hydrolyzed to gluconate

- Need to open ring

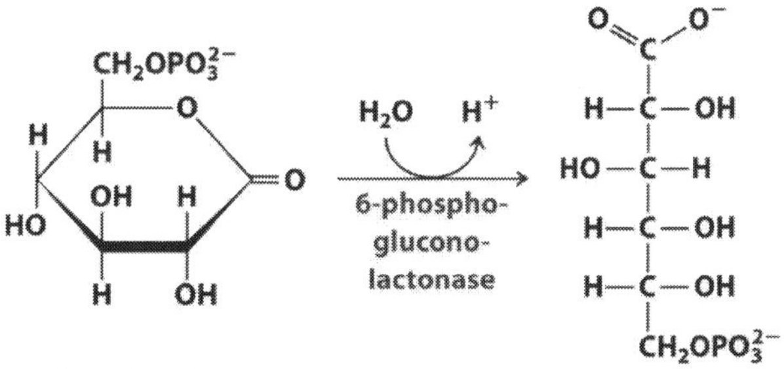

6-Phosphoglucono-δ-lactone → 6-Phosphogluconate

6-phospho-glucono-lactonase

Step 3

- 6-phosphogluconate oxidized and decarboxylated to give a 5 carbon sugar

- Reduces $NADP^+$ to NADPH

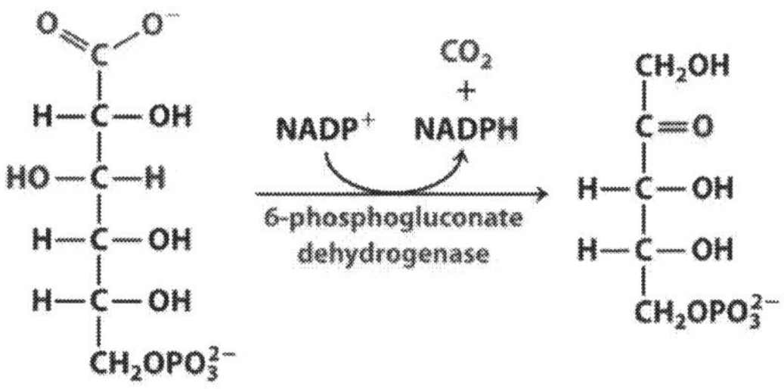

6-Phosphogluconate → Ribulose-5-phosphate

6-phosphogluconate dehydrogenase

Step 4

- Ribose-5-P formed

- Convert ketose to aldose

- Needed to form proper ring structure in product

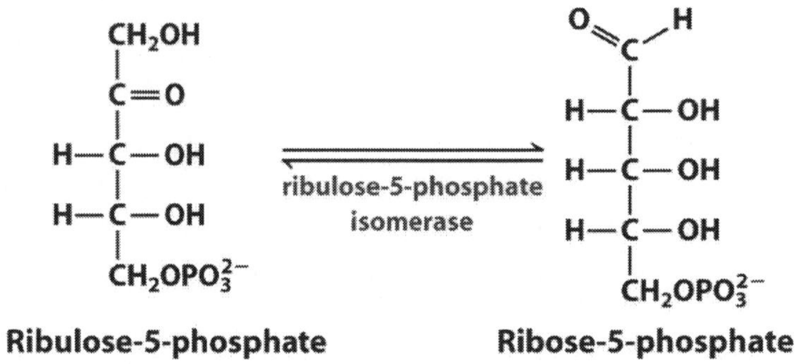

Ribulose-5-phosphate ⇌ (ribulose-5-phosphate isomerase) **Ribose-5-phosphate**

- Pathway activity is high in rapidly dividing cells

  o Require large amounts of DNA

- NADPH can be used to reduce Rib-5-P to dRib-5-P (deoxy)

  o Using both main products of pathway

- Cells may need NADPH more than they need Rib-5-P

  o Excess carbons recycled to build glycolytic intermediates

  o Sugars can then be degraded to pyruvate or used in gluconeogenesis

  o Readily reversible reactions

    ▪ Interconvert sugars

CHAPTER 13: THE CITRIC ACID CYCLE

Citric Acid Cycle

The citric acid cycle is also known as the TCA (tricarboxylic acid) cycle and is also known as the Krebs cycle. It is a series of chemical reactions used by all aerobic organisms to generate energy.

- Sequence of 8 catalyzed reactions

- Acetyl group enters pathway

- For each acetyl group two molecules of CO_2 is produced

  - Electrons are passed to 3 NAD^+ and 1 Q (ubiquionone)

- Acetyl-CoA + GDP + P_i + 3 NAD^+ + Q $\rightarrow$ 2 CO_2 + CoA + GTP + 3 NADH + QH_2

Step 1: Formation of Citrate

- Carried out by citrate synthase

- Breakage of thioester bond

- Add acetyl to carbonyl carbon

- Convert 4C to 6C compound

- $\Delta G^{\circ\prime}$ = -31.5 kJ/mol

Oxaloacetate   Acetyl-CoA   Citrate

Step 2: Isomerization of Citrate to Isocitrate

- Catalyzed by acontinase

- Move –OH from C_3 to C_4

- Need to form C=O at C_4

- Reversible Reaction

The reaction scheme shows the conversion of Citrate to Aconitate (in brackets) with H_2O, then Aconitate to Isocitrate with H_2O, both reversible.

Citrate ⇌ **Aconitate** ⇌ **Isocitrate**

Step 3: Oxidative Decarboxylation of Isocitrate to Form α-Ketoglutarate

- Catalyzed by isocitrate dehydrogenase
- Oxidized to form keto acid; NAD^+ reduced (energy stored as NADH)
- Carboxylate group eliminated as CO_2
- Mn^{2+} in active site stabilizes negative charges on intermediate
- Irreversible reaction

The reaction scheme shows Isocitrate converting with $NAD^+ \rightarrow H^+ + NADH$, then losing CO_2, then gaining H^+ to form α-Ketoglutarate.

Isocitrate → → → **α-Ketoglutarate**

Step 4: Oxidative Decarboxylation to Form Succinyl-CoA

- Catalyzed by α-ketoglutarate dehydrogenase
- Step 3 and 4 release 2 CO_2 molecules
  - Not the carbon atoms that entered as acetyl-CoA
- Net effect: lose 2 C atoms for every acetyl-CoA that enters cycle

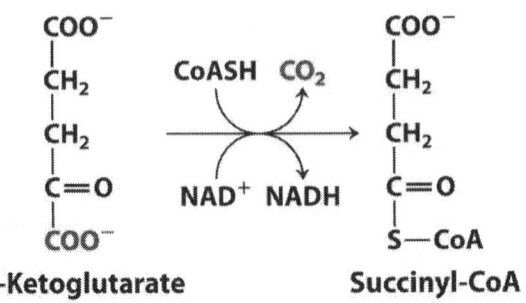

α-Ketoglutarate → (CoASH, CO_2 released; $NAD^+ \rightarrow$ NADH) → **Succinyl-CoA**

Step 5

- Catalyzed by succinyl-CoA synthetase

- Driven by cleavage of thioester bond

- A phosphate group displaces CoA in succinyl-CoA, and produces succinyl phosphate

    o Succinyl phosphate is an acyl phosphate

        ▪ Releases a large amount of free energy when cleaved

- Succinyl phosphate donates its phosphoryl group to a *His* residue on the enzyme

    o Produces a phosphor-*His* intermediate and releases succinate

- Phosphoryl group is then transferred to GDP to form GTP

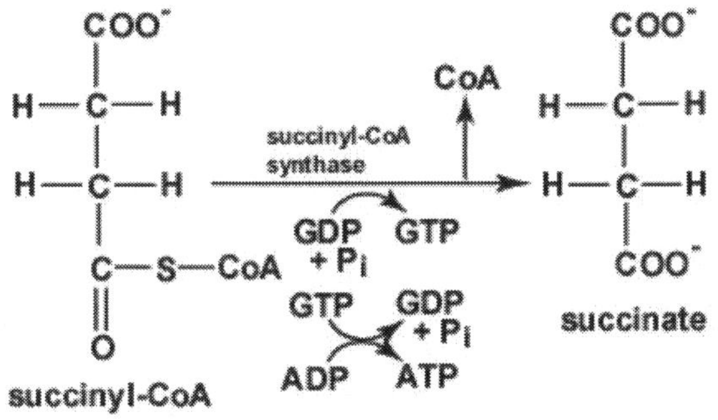

Step 6: Oxidation of Succinate to Fumarate

- Catalyzed by succinate dehydrogenase

- Formation of a C=C bond

- Electrons passed to FAD enzyme cofactor to form $FADH_2$

- Need to generate FAD

- Enzyme in mitochondrial membrane – need lipid-soluble electron carrier

    o Electrons passed to ubiquinone (Q) to form ubiquinol (QH_2)

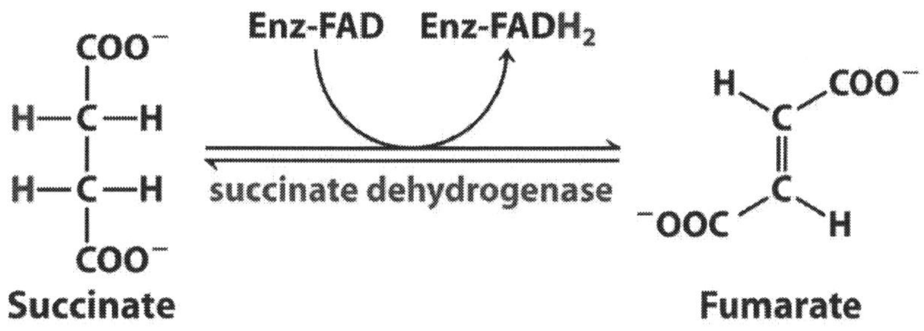

Step 7: Hydration of Fumarate to Malate

- Catalyzed by fumarase

- Need O atom to ultimately form C=O bond in oxaloacetate

$$\text{Fumarate} \quad \xrightarrow[\text{fumarase}]{H_2O} \quad \text{Malate}$$

Step 8: Oxidation of Malate to Oxaloacetate

- Catalyzed by malate dehydrogenase

- Electrons passed to NADH (energy recovered)

  o $\Delta G^{\circ'} = +29.7$ kJ/mol

- Product is substrate for citrate synthase

  o $\Delta G^{\circ'} = -31.5$ kJ/mol

  o This is what pulls step 8 forward

$$\text{Malate} \quad \xrightarrow[\text{malate dehydrogenase}]{NAD^+ \quad NADH + H^+} \quad \text{Oxaloacetate}$$

Regulation of the Citric Acid Cycle

The Citric Acid Cycle is regulated at irreversible steps.

- Step 1
    - Flux forward depends largely on substrate concentrations
    - Inhibited by citrate and succinyl-CoA (feedback inhibition) and NADH
- Step 3
    - Inhibited by NADH
    - Activated by Ca^{2+} and ADP
        - Need for fuel
        - Influx of Ca^{2+} in response to epinephrine
- Step 4
    - Inhibited by succinyl-CoA and NADH
    - Activated by Ca^{2+}

Intermediates as Precursors

The Citric Acid Cycle is a central pathway that provides key intermediates that feed into other pathways to produce necessary metabolites.

- Intermediates are precursors for other anabolic pathways
    - Need to replenish the intermediates that are siphoned off
- Anaplerotic reactions – reactions that form intermediates of a metabolic pathway

CHAPTER 14: OXIDATIVE PHOSPHORYLATION

What is Oxidative Phosphorylation?

Oxidative phosphorylation (OXPHOS) is a metabolic pathway in which mitochondria use the energy released by the oxidation of nutrients to reform ATP.

- Oxidative

    - Process that involves electron extraction

    - Electrons are passed through the electron transport system and ultimately to a terminal electron acceptor

- Phosphorylation

    - Synthesis of ATP from ADP + P_i

    - Converts energy stored as a proton gradient into energy stored in a chemical bond

- Most molecules of ATP are produced by OXPHOS

    - Only 2 net ATPs generated for each glucose molecule that goes through glycolysis

    - Only 2 ATPs per glucose are produced from the Krebs Cycles

- Most of the energy extracted from glucose is in molecules of NADH and $FADH_2$ at the end of the Krebs Cycle

    - The reduced coenzymes link glycolysis and the Krebs cycle to oxidative phosphorylation by passing their electrons down the ETC (electron transfer chain) to oxygen

    - ETC and oxidative phosphorylation require oxygen as the final electron acceptor

Redox Thermodynamics

If you extract electrons from one donor, you must deposit them on an acceptor.

- $FADH_2 + Q \longleftrightarrow FAD + QH_2$

- Reaction can be expressed as half-reactions

    - Only look at one substance at a time

    - $Q + 2 H^+ + 2 e^- \longleftrightarrow QH_2$

- Standard reduction potential (E°')

$$E = E^{\circ\prime} - \frac{RT}{nF} \ln \frac{\left[A_{reduced} \right]}{\left[A_{oxidized} \right]}$$

  o Is the tendency of the oxidized form of a substance to be reduced

  o The more positive the value, the greater the tendency to be reduced

  o Actual reduction potential depends on the concentrations of the substances involved

  o Reduction potential predicts the flow of electrons

    ▪ Electrons flow from a substance with lower potential to a substance with higher potential

    ▪ Lower or negative reduction potentials mean that a substance is more likely to donate than receive electrons

Free Energy Charges

- The larger the positive value of ΔE, the greater the tendency of the electrons to flow in that direction and the greater the change in free energy

- Large negative values of ΔG requires a positive value for ΔE

- Electrons flow down the transport chain according to reduction potential

Mitochondrion

Electron transport takes place in the mitochondrion.

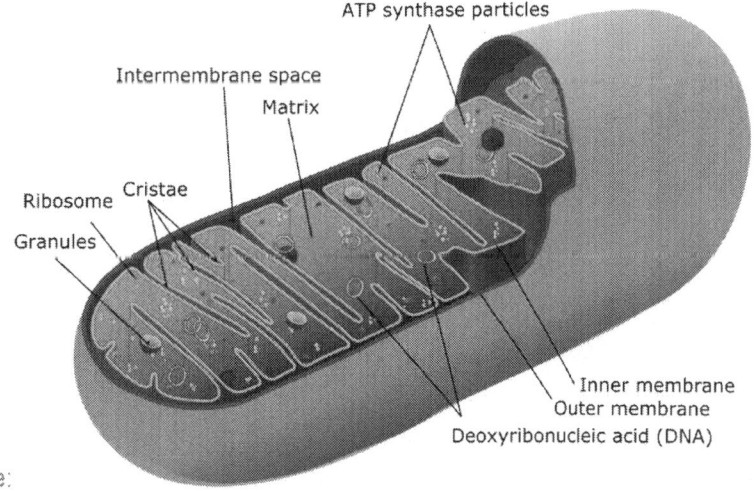

Mitochondrion Structure:

- Cristae – folds of inner membrane

    o Highly variable in structure

- Outer membrane

    o Permeable to small molecules and ions

- Inner Membrane

    o Impermeable to most small molecules

    o Vital for forming and maintaining proton gradient

    o Carries ETC, ATP synthase, and translocase to move ADP in and ATP out

- Matrix

    o Contains oxidation enzymes

        ▪ Except glycolytic enzymes

    o Also contains DNA, ribosomes, other enzymes, metabolic intermediates, small molecules, and ions

    o Inner membrane keeps it separate from cytosolic components

- Intermembrane space

    o Area between the inner and outer membranes

Transport Shuttle Systems

Reducing equivalents need to be moved from the cytosol to the matrix.

- Electrons are donated to the electron transport chain on the matrix side

- Transfer electrons to molecule that can be transported across membrane

- Then the molecule is reoxidized to extract the electrons

Electron Transport

- Electrons extracted during catabolic oxidation are passed to universal acceptors like NAD^+ or $NADP^+$ (diffusible carriers)

    o Passed sequentially through a series of membrane-bound carriers

- Energy is used to pump protons from the matrix to the intermembrane space

- Electrons are eventually passed to a terminal acceptor

Electron Transport Systems

Electron transport systems (ETS) are series of electron acceptors.

- The purpose of the ETC is to convert some of the chemical energy extracted from food to a form that can be used to make ATP

- ETC is composed of electron-carrier molecules built into the inner mitochondrial membrane

- They accept energy-rich electrons from reduced coenzymes

    - NADH and $FADH_2$

- Following a series of redox reactions, they pass down electrons down the chain to oxygen, the final electron acceptor, and the molecule with the highest electronegativity in the chain

- The electronegative oxygen accepts these electrons, along with H^+ to form water

    - Electron transfer from NADH to oxygen is exergonic

        - Free energy change of -222 kJ/mole (-53 kcal/mol)

- Because electrons lose potential energy when they shift toward a more electronegative atom, these redox reactions releases energy

Mitochondrial Respiratory ETS - Summary

- 4 Complexes

    - I: Electrons passed from NADH to Coenzyme Q

    - II: Electrons passed from $FADH_2$ to Coenzyme Q

    - III: Electrons passed from Q to Cytochrome c

    - IV: Electrons passed from Cytochrome c to O_2

- Alternative steps drive protons out of the matrix

Complex I

- Electrons transfer from NADH to FMN, then to Fe-s and finally to Q

- As electrons are transferred from NADH to ubiquinone (Q), Complex I transfers four protons from the matrix to the intermembrane space

    - The 4 H^+ are relayed through hydrogen bonding network

Complex III

- Complex III transfers e- from QH_2 to cytochrome c

- Aka: cytochrome bc_1

Heme b

- Cytochromes are proteins with heme groups

- Letters (a ,b, c, etc.) identify the structure of porphyrin ring of heme

- Unlike heme groups in Hb and Mb, cytochrome c heme undergoes reversible one-electron transfers

- The central iron atom is either oxidized (Fe^{3+}) or reduced (Fe^{2+})

Mammalian Complex III

- Contains 2 cytochromes (b and c_1) with heme groups + Fe-S protein

   - Prosthetic groups arranged so electrons flow from matrix to intermembrane space

- Net reaction: QH_2 + 2 cytochrome c (Fe^{3+}) $\longleftrightarrow$ 2 cytochrome c (Fe^{2+})

- Transfer 2 e- from QH_2 to 2 cytochrome c proteins

Q Cycle

Flow of e- through Complex III:

- QH_2 donates e- to Fe-S protein (ISP)

   - e- travels to cytochrome c_1 then to cytochrome c

- QH_2 donates e- to cytochrome b

   - Cytochrome b and ISP can't accept H^+

   - 2 H^+ released into intermembrane space

- Q diffuses to 2nd binding site
  - Picks up e^- from cytochrome b
  - It becomes: $\cdot Q^-$ (semoquinone)
  - Donating and then picking up the same e^- pumps protons
- 2nd molecule of QH_2 donates e- to Fe-S protein (ISP)
  - e^- travels to cytochrome c_1 then to cytochrome c
- Same QH_2 then donates e- to cytochrome b
  - 2 H^+ released into intermembrane space (4 H^+ total)
- $\cdot Q^-$ picks up e^- of round 2 + $2H^+$ from matrix to become QH_2
- Net effect: 2 QH_2 → 1 QH_2 + Q
  - This is why >1 QH_2 is needed in the Q pool

Cytochrome c

- Cytochrome c is a small soluble protein in the intermembrane space
- Cytochrome c transfers one electron at a time from Complex III to Complex IV
- Complex IV (cytochrome oxidase) catalyzes the reaction:
  - 4 cytochrome c (Fe^{2+}) + O_2 + 4 H^+ → 4 cytochrome c (Fe^{3+}) + 2 H_2O

Complex IV Function

- For every two electrons donated by cytochrome c_1 two protons are translocated to the intermembrane space
- Only Complex IV has Cu^{2+} redox centers
- Two protons from the matrix are also consumed in the reaction: $\frac{1}{2} O_2$ → H_2O

Chemiosmotic Model

- Energy stored in proton "battery" used to drive ATP synthesis
- Flow of protons back into matrix provides energy to synthesize ATP
  - ADP + P_i + nH^+_p → ATP + H_2O + nH^+_n

- Proton Motive Force
    - The imbalance of protons represents a source of free energy, also called a proton motive force, that can drive the activity of an ATP synthase
- Free Energy Change of H$^+$ Gradient

$$\Delta G = RT \ \text{In} \ \frac{[H^+_{out}]}{[H^+_{in}]} + ZF\Delta\psi$$

    - Z = Ion's Charge
    - F = Faraday's Constant
    - $\Delta\Psi$ = Membrane Potential

ATP Synthase

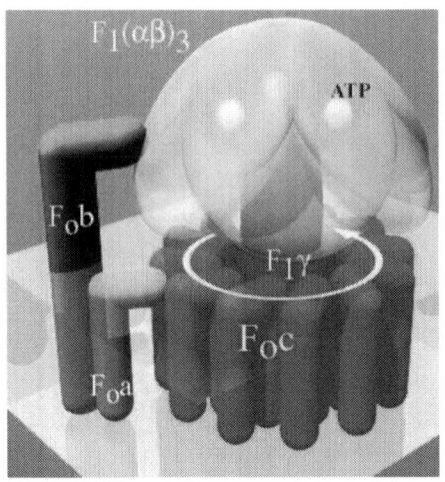

- Also called F$_o$F$_1$ ATP synthase
    - F$_o$ – sensitive to oligomycin; proton pore
    - F$_1$ = where ATP synthesis takes place
- Also called Complex V
- F$_o$ cylinder composed of 10 c subunits embedded in membrane
- Attached to F$_1$ shaft of γ and ε
- 2 b subunits of F$_o$ hold α and β fixed relative to membrane
- Nucleotide binding sites are in β subunits, near interfaces with α
    - One binding site has ATP, one has ADP, and one has none

- - 3 identical β subunits, yet their conformations are not the same
- Rotational Catalysis
  - Protons flow through the F_o cylinder
  - Causes cylinder and therefore γ/ε shaft, to rotate
  - As the shaft rotates 120°, γ comes in contact with a different β subunit
  - Conformations change from:
    - B-Loose (ADP bound)
    - To β-Tight (high ATP affinity)
    - To β-Open (low ATP affinity)
  - 3 H^+ pumped for each ATP made
- Isolated F_1 catalyzes ATP hydrolysis
  - Called F_1 ATPase
  - $\Delta G^{\circ}{}' = -30.5$ kJ/mol

P:O ratio

# of phosphorylations of ADP (P) relative to # of O_2 reduced (O).

- Oxidation of NADH
  - Pump 10 H^+ for each ½ O reduced
  - Mitochondrial ATP synthase has 8 c subunits in F_o
    - 1 H^+ pumped through each c subunit
  - 1 full rotation of F_o: 8 H^+ per 3 ATP or 2.7 H^+ per 1 ATP
  - P:O ratio ~3 (10 H^+ pumped/ 2.7 = 3.7)
- Oxidation of QH_2
  - Pump 6 H^+ for each ½ O_2 reduced
  - P:O = 2.2
- Never a whole integer (even # H^+ pumped per odd # ATP made)
- Actual P:O ratio lower in vivo

CHAPTER 15: PHOTOSYNTHESIS

Chloroplast

The thylakoid membrane inside a chloroplast is where photosynthesis takes place.

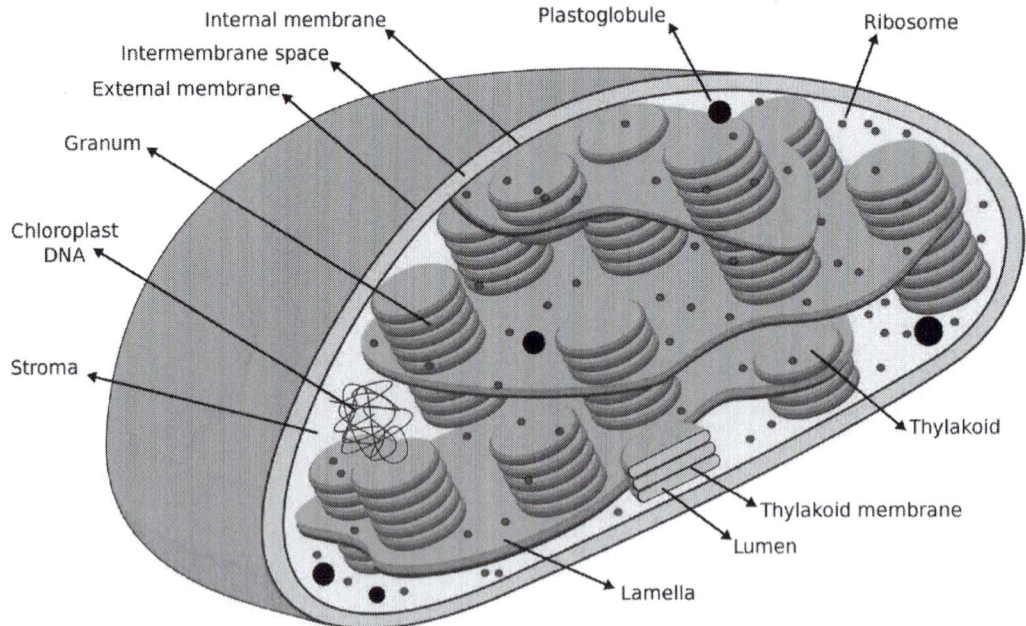

Energy of Photons and Light Pigments

Energy of a photon

- E= hc/λ

  - h is the Planck's constant = 6.626 X 10^{-34} joules•sec

  - c is the speed of light in a vacuum = 2.998 X 10^8 m/sec

- Shorter λ = higher energy

Light Pigments

- Photons are absorbed by pigments or photoreceptors

  - Chlorophyll

  - Carotenoids

  - Phycocyanin

- Different photoreceptors absorb different wavelengths

Fates of Absorbed Energy

Absorption of photon promotes delocalized electron to an excited state (higher-energy orbital).

- Electron in the excited state returns to ground state by four mechanisms
    - Lost as heat
    - Emitted as light: fluorescence
        - Emitted light has less energy (longer λ)
    - Exciton transfer: energy transferred to another molecule
    - Photooxidation: electron itself transferred

Light-Harvesting Complex

The protein environment of a photoreceptor influences the λ of light that is absorbed.

- Reaction centers are where photosynthesis takes place
    - Contain chlorophyll molecules
- Light-harvesting complexes
    - Antenna pigments in membrane proteins
    - Consist of many different pigment types each absorbing a different λ of light
    - Exciton transfer brings energy to chlorophyll at a reaction center
- Harvesting Light
    - Photoreceptor that absorbs higher energy (shorter λ) and transfers energy to one that absorbs lower energy (longer λ)

Light Reactions

The light reactions convert energy from the sun into chemical energy in the form of NADPH and ATP. Noncyclic flow and cyclic flow are two possible routes of electron flow during light reactions.

Result of light reactions

- Water oxidized to O_2
- $NADP^+$ reduced to NADPH

- H^+ gradient produced

- Redox centers similar to those in ETS of mitochondria

Photosystem II

Photosystem II is the first protein complex in the light-dependent reaction of photosynthesis

- Antenna assembly of photosystem II absorbs a photon

  - Energy is transferred to the P680 reaction center

- Excited state P680* reaction center quickly gives up the electron to the primary electron acceptor

- From the primary electron acceptor, electrons are transferred to an electron transport chain in the thylakoid membrane

  - Plastoquinone (PQ) is the first carrier in the chain

  - Through a series of redox reactions the electrons travel from PQ → complex of two cytochromes → plastocyanin (PC) → P700 of photosystem II

    - Electrons flow spontaneously from low to high reduction potentials

- Electrons from P680 flow to P700 during noncyclic electron flow

  - This leaves the P680 reaction center of photosystem II with missing electrons

  - The electrons needed to reduce $P680^+$ come from water

- A water splitting enzyme extracts electrons from water and passes them to the oxidized P680 chlorophyll

  - As water is oxidized, it splits off an oxygen atom which immediately combines with a second oxygen atom to form O_2

    - This step is what generates most of the atmospheric oxygen

- As excited electrons give up energy along the transport chain to P700, the thylakoid membrane couples the exergonic flow of electrons to the endergonic reactions that phosphorylate ADP to ATP

  - Coupling mechanism is chemiosmosis

- Some electron carriers can only transport electrons with protons

    o The protons are picked up on one side of the thylakoid membrane and deposited on the opposite side

        - This is how electron flow stores energy in the form of a proton gradient across the thylakoid membrane (proton-motive force)

    o An ATP synthase enzyme in the thylakoid membrane uses the proton-motive force to make ATP

        - This process is called photophosphorylation because the energy required comes from light

        - This form of ATP production is called noncyclic photophosphorylation

Photosystem I

- Light excites electrons from P700

- The excited state electrons are transferred from P700 to the primary electron acceptor of photosystem I

- The primary electron acceptor then passes these electrons to ferredoxin (a small peripheral protein on stromal side)

- There are 2 different fates for electrons gathered by ferredoxin:

    o Noncyclic electron flow

    o Cyclic electron flow

Noncyclic Electron Flow

- $NADP^+$ reductase catalyzes the redox reaction that transfers these electrons from ferredoxin to $NADP^+$ producing NADPH

Cyclic Electron Flow

- Cyclic electron flow involves only photosystem I and generates ATP without producing NADPH or evolving oxygen

- Cyclic electron flow is cyclic because excited electrons that leave from chlorophyll *a* at the reaction center return back to the reaction center

- From ferredoxin the excited state electrons take an alternate path that sends them down the electron transport chain to P700

    o This is the same electron transport chain involved in noncyclic electron flow

- The exergonic flow of electrons is coupled to ATP production by chemiosmosis
    - This process of ATP production is called cyclic photophosphorylation
- Absorption of another two photons of light by the pigments sends a second pair of electrons through the cyclic pathway
- The function of the cyclic pathway is to produce additional ATP
    - This is accomplished without the production of NADPH or O_2
- Benefits of cyclic electron flow
    - Cyclic photophosphorylation provides another ATP source required for the Calvin cycle and other metabolic pathways
    - The noncyclic pathway produces approximately equal amounts of ATP and NADPH
        - Useful when there may not be enough ATP to meet demand
        - Cyclic electron flow allows for higher ATP production when needed

Summary

- During noncyclic electron flow, the photosystems of the thylakoid membrane transform light energy to the chemical energy stored in NADPH and ATP
    - Pushes low energy-state electrons from water to NADPH
    - NADPH can then be used as the electron donor used to reduce carbon dioxide to sugar (Calvin cycle)
    - Produces ATP
    - Produces oxygen as a by-product
- During cyclic electron flow, electrons ejected from P700 reach ferredoxin and flow back to P700
    - Produces ATP
    - Unlike noncyclic electron flow, does not produce NADPH or O_2

Calvin Cycle

The Calvin Cycle is also known as the light-independent reactions or the "dark reactions." ATP and NADPH produced by the light reactions are used in the Calvin cycle to reduce carbon dioxide to sugar (the 3-carbon sugar glyceraldehyde-3-phosphate). ATP is the energy source, while NADPH is the reducing agent that adds high energy electrons to form sugar.

- 3 phases
  - Carbon fixation
  - Reduction
  - Regeneration

Phase 1: Carbon Fixation (Rubisco Reaction)

- The Calvin cycle begins when each molecule of CO_2 is attached to a five-carbon sugar, ribulose biphosphate (RuBP).
  - Catalyzed by RuBisCO
    - Ribulose bisphosphate carboxylase/oxygenase
    - Most abundant biological catalyst
    - ½ of protein content of chloroplast
    - Can act as either carboxylase or oxygenase
- The product of this reaction is an unstable six-carbon intermediate
  - It immediately splits into two molecules of 3-phosphoglycerate
- For every three CO_2 molecules that enter the Calvin cycle, three RuBP molecules are carboxylated forming six molecules of 3-phosphoglycerate

Phase 2: Reduction

- A two-step process that couples ATP hydrolysis with the reduction of 3-phosphoglycerate to glyceraldehyde phosphate
- An enzyme phosphorylates 3-phosphoglycerate by transferring a phosphate group from ATP
  - Produces 1, 3-bisphosphoglycerate
  - Six ATP molecules are used to produce six molecules of 1,3-bisphosphoglycerate

- Electrons from NADPH reduce the carboxyl group of 1,3-bisphosphoglycerate to the aldehyde group of glyceraldehyde 3-phosphate (G3P)

    o The product, G3P, stores more potential energy than the initial reactant, 3-phosphoglycerate

    o G3P is the same three-carbon sugar produced when glycolysis splits glucose

- For every three CO_2 molecules that enter the Calvin cycle, six G3P molecules are produced (net gain is only one)

    o The cycle begins with three five-carbon RuBP molecules (total of 15 carbons)

    o The six G3P molecules produced contain 18 carbons

        ▪ Net gain of 3 carbons that come from CO_2

    o One G3P molecule exits the cycle

        ▪ Other five are recycled to regenerate three molecules of RuBP

Phase 3: Regeneration

- A complex series of reactions rearranges the carbon skeletons of five G3P molecules into three RuBP molecules

    o Net Reaction: $5 C_3 \rightarrow 3 C_5$

    o These reactions require three ATP molecules

    o RuBP is regenerated to start the cycle again

- For the net synthesis of one G3P molecule, the Calvin cycle uses products of the light reactions:

    o 9 ATP molecules

    o 6 NADPH molecules

- Uses of Gly-3-P

    o Synthesis of glucose and amino acids

    o Converted to pyruvate then to oxaloacetate – used to make other amino acids

- The Calvin cycle uses 18 ATP and 12 NADPH molecules to produce one glucose molecule

    o This is an expensive process

- Net reaction:

    - $3\ CO_2 + 9\ ATP + 6\ NADPH \rightarrow Gly\text{-}3\text{-}P + 9\ ADP + 8\ P_i + 6\ NADP^+$

    - To fix $1\ CO_2 \rightarrow 3\ ATP + 2\ NADPH \rightarrow$ generated from about 8 photons

    - To make 1 glucose $\rightarrow$ 18 ATP and NADPH

Photorespiration

Photorespiration is a metabolic pathway in plants that consumes oxygen, evolves carbon dioxide, produces no ATP and decreases photosynthetic output. Photorespiration is a way to dissipate free energy (ATP + NADPH) from light reactions when there is not enough CO_2 for carbon fixation.

- Occurs because the active site of rubisco can accept both O_2 and CO_2

- When the O_2 concentration is higher than CO_2 concentration

    - Rubisco accepts O_2 and transfers it to RuBP

- Photorespiration is promoted by hot, dry, and sunny days

    - Plants close their stomata to prevent dehydration

    - Photosynthesis then depletes CO_2 and increases oxygen

        - This condition favors photorespiration

Regulation of Calvin Cycle

Availability of light is the key in regulation of the Calvin Cycle.

- Photosynthesis and Calvin cycle operate during day

- At night, "dark reactions" turned off to conserve ATP and NADPH

    - Turn on pathways to regenerate these (glycolysis; pentose phosphate)

    - Dark reactions don't occur in the dark – the name is a misnomer

- High pH favors carboxylation of Lys residue on RuBisCO

    - Signal that the light reactions are pumping protons out of stroma and that ATP and NADPH are available for Calvin cycle

Storing Carbohydrates

The 3-C sugars produced by Calvin cycle are converted to sucrose or starch.

- 2 Gly-3-P → Glc-6-P (similar to gluconeogenesis)

- Glc-6-P → Glc-1-P

- Sugar activated by AMP transfer

  - Similar to activation to form glycogen in mammals

  - Driven by PP_i hydrolysis

- Starch synthase transfers glucose to end of starch polymer

- 3-C converted to 6-C sugars

  - UDP-Glucose and Fructose-6-phosphate

- Fru-6-P reacts with UDP-Glc to form sucrose-6-P

- Then converted to sucrose

  - Transport form of carbon in plants

  - Linkage not subject to amylases or other hydrolases

    - Very stable

CHAPTER 16: LIPID METABOLISM

Major Pathways of Lipid Metabolism

- Fatty acid catabolism (β oxidation)

- Fatty acid synthesis

- Ketogenesis

    - Synthesis of ketone bodies

- Cholesterol synthesis

- Synthesis of other lipids

- All Pathways converge at acetyl-CoA

Overview of Lipid Metabolism

- Fat storage is the most important form of energy storage

- 3 sources

    - Dietary triacylglycerols (TAGs)

    - TAGs synthesized in liver

    - TAGs stored in adipocytes

- Many diseases are associated with disruptions in lipid metabolism

Atherosclerosis

Atherosclerosis is a slow progressive disease, characterized by the hardening of the arteries due to lipid accumulation in blood vessel walls.

- 50% of all deaths in U.S. linked to this vascular disease

- Inflammation initiated and perpetuated by leukocytes (white blood cells)

- Plaque forms, containing cholesterol, dead macrophages, and calcified muscle

- Rupture of plaque releases blood clot which can lead to:

    - Heart attack – disruption of circulation to heart

    - Stroke – disruption of circulation to brain

Lipoproteins

Lipoproteins contain both a protein component and a lipid component.

- They are the primary source of circulating lipids

- Protein component makes them soluble

Types of lipoproteins

- **Chylomicrons**

    o Dietary lipids that travel from intestine to other tissues

    o Large (1000 - 5000 Å) and mostly lipid (1-2% protein)

    o Transport triacylglycerols to adipose tissue and cholesterol to liver

- **VLDL (very-low-density lipoprotein)**

    o Particles of TAGs, phospholipids, cholesteryl esters packaged by liver

    o Circulate in blood and donate some of TAG to tissues

    o Become smaller and denser and richer in cholesterol as they donate TAGs

- **IDL (intermediate-density lipoproteins)**

    o Form as VLDL gives off TAG

    o Intermediate state between VLDL and LDL

- **LDL (low-density lipoprotein)**

    o This is known as the "bad cholesterol"

    o Form as more TAG is given off from IDL particles

    o High levels correlated to atherosclerosis

    o Other factors involved in disease

- **HDL (high-density lipoprotein)**

    o This is known as the "good cholesterol"

    o Primary function is to transport excess cholesterol back to liver

    o Remove excess cholesterol from cells

- Lipids Transported via Lipoproteins
    - As % protein ↑ density ↑ and diameter ↓
    - Lipoproteins transport cholesterol and other fats

Fatty Acid Catabolism

Degradation of fatty acids (FA) is known as β oxidation.

- Dietary TAGs are the primary source in humans
- TAGs are hydrolyzed to FA groups by lipases
- FA in bloodstream bound to albumin
    - ~50% of the protein in the blood is albumin
    - Not water-soluble

Activation of Fatty Acids

Fatty acids must be activated before they can be oxidized.

- Goal is to form acyl-CoA (acyl-CoA synthetase)
- Breaking a thioester bonds releases a lot of free energy
    - It takes the input of large amounts of energy to form one
- Energy comes from phosphoanhydride bond of ATP
- What drives the reaction and makes it irreversible?
    - Hydrolysis of PP_i
- Energy cost: 2 ATP-equivalents
    - 2 phosphoanhydride bonds

β -Oxidation

β-Oxiation is a spiral process.

- Degrades acyl-CoA to produce acetyl-CoA (C_2 compound)
- Each round shortens acyl chain by 2 C atoms
- For a C_{16} fatty acid; 7 rounds
    - 7 bonds connecting 8 C_2 groups

- Net products of each round
    - 1 QH_2
    - 1 NADH
    - 1 acetyl-CoA
        - TCA: 3 NADH + 1 QH_2 + 1 GTP
    - Total ATP Yield = 14

Steps of β-Oxidation

- Want to remove acetyl group on end and create new acetyl group on the remaining acyl chain
    - Oxidize to produce C=C bond between C_2 and C_3
        - 2 e⁻ passed to FAD; then Q
        - *Trans* bond
    - Add water across bond
    - Oxidize to form carbonyl
    - Hydrolyze to remove acetyl-CoA; need to add new CoA to end of remaining acyl chain
- Repeat until chain completely degraded to acetyl-CoA units

Unsaturated Fatty Acid Catabolism

Double bonds in fatty acyl chains are cis bonds.

Example: Linoleate

- Steps proceed normally until round 4
- Double bond present already so don't need step 1
    - But hydratase needs *trans* configuration to carry out its function
    - Isomerase makes the switch
        - Back to normal process
- Problem again with round 5 from step 1
    - Now have dienoyl
        - Not a good substrate for step 2 enzyme

- ○ Reductase converts it to a single *trans* 3,4 double bond
- The interruptions in the typical cycle net less ATP
  - ○ Bypass QH_2 production of step 1
  - ○ Consume NADPH that could contribute to ETS (energy equivalent of NADH)

Oxidation of Odd-Chain Fatty Acids

Humans make even numbered FAs, however, some plants and bacteria make odd chain FAs and we ingest them.

- Final round leaves propionyl-CoA (3-C fragment)
  - ○ Takes 8 steps to make acetyl-CoA
    - Carboxylase adds CO_2 to C_2
      - Not chemically possible to add it to C_3
      - Need adjacent carbonyl
    - Racemase converts to only (R) isomer
      - R vs. S based on orientation of chiral center by priority rules
    - Mustase moves COO- to C_3 carbon to form Succinyl-CoA
    - Steps 4-6 are those that are part of the TCA cycle
    - Step 7: oxidize to form C=O and remove CO2
    - Step 8: oxidative decarboxylation
      - Remove CO2
      - Add CoA
      - Enters TCA cycle as substrate at step 1

Beta-Oxidation in Peroxisomes

- Membrane-bound organelle in most eukaryotic cells
- Site of β-oxidation of long chain fatty acid
  - Differs in 1st step
    - Still an oxidation using FAD prosthetic group
    - Electrons passed to O_2 directly, rather than to Q
    - Peroxide product gives organelle its name
  - This is a chain-shortening system
  - Shorter chains can then move to mitochondria for complete oxidation
  - Peroxisomes also contain enzymes to oxidize branched-chain fatty acids

Fatty Acid Synthesis

Pathways of fatty acid degradation and synthesis are spatially separate. Oxidation takes place in the mitochondrial matrix and synthesis takes places in the cytosol.

- Different cofactors used
  - Acyl chain bound to CoA in oxidation
  - Bound to acyl carrier protein (ACP) in synthesis
- Different electron acceptors/donors
  - Oxidation passes electrons to NAD+ and Q (Catabolic)
  - NADPH is electron donor in synthesis (anabolic)
- Differ in ATP requirements
  - 2 ATP needed to activate acyl group once for multiple rounds of oxidation
  - 1 ATP consumed for each acetyl added to synthesize acyl chain

Substrate Transport

- Acetyl-CoA is the starting material for FA synthesis
- Needs to exit into the cytosol

- Acetyl transported as citrate
  - Acetyl-CoA converted to citrate via citrate synthase
    - Step 1 of TCA cycle
  - Cytosolic ATP-citrate lyase converts it back
    - ATP hydrolysis drives formation of thioester bond

Acetyl-CoA Carboxylase

- First biotin is carboxylated
  - Biotin + HCO_3^- + ATP → biotin-COO^- + ADP + P_i
- Second, carboxyl group is transferred to acetyl-CoA

Fatty Acid Synthase

- 1 gene encodes 1 polypeptide that is actually 7 enzymes
- Protein is a homodimer with 6 active sites per polypeptide
- Product of one reaction quickly diffuses to the next active site

Fatty Acid Synthesis Reactions

- **Reactions 1 and 2 load the enzyme with substrate**
  - Acetyl group loaded onto Cys residue
  - Malonyl transferred from CoA to ACP domain of enzyme
  - Both have thioester link to the enzyme
- **Reaction 3 is a condensation reaction**
  - Decarboxylation of malonyl allows C_2 to attack carbonyl of acetyl group
    - This is why the CO_2 group was added in the first step
- **Reaction 4 is a reduction that converts carbonyl to hydroxyl**
- **Reaction 5 is a dehydration, removes water to form C=C bond**
- **Reaction 6 is a reduction that fully saturates the chain**

- **Reaction 7 transfers chain to Cys residue**
    - Another malonyl group loaded onto enzyme
    - Start process again
- Process continues to the formation of palmitoyl-ACP
- Thioesterase releases product
- NADPH mostly comes from Pentose Phosphate Pathway
- Cost of fatty acid synthesis (palmitate example):
    - 7 malonyl-CoA = 7 ATP
    - 14 NADPH (14 X 2.5 = 35 ATP)
    - Total cost: 42 ATP
    - Full oxidation of palmitate yields more than it costs to make it

Generating Longer Chains

- Elongases extend chains beyond C_{16}
- Desaturases introduce double bonds
- Different combinations of elongations and desaturations produce chains of varying lengths and degrees of saturation
- Mammals cannot introduce double bond beyond C_9
    - These are the essential fatty acids
        - Must be supplied through our diet

Regulation of Fatty Acid Synthesis

Rate of synthesis is controlled at the first step.

- Acetyl-CoA carboxylase
    - Inhibited by palmitoyl-CoA
        - Feedback inhibition
- Activated by citrate
    - Indicates there is plenty of acetyl-CoA
- Allosterically regulated by +/- phosphoryl group in response to hormone signals

- Malonyl-CoA blocks beta-oxidation
  - Inhibits carnitine acyltransferase

Ketogenesis

- Fatty acids are vital energy source during fast
  - Glucose is unavailable and glycogen is depleted in fasted state
- Brain cannot burn FAs for fuel
  - Passes poorly through blood-brain barrier
  - Gluconeogenesis is important
- Liver also produces ketone bodies from acetyl-CoA in mitochondria
- 3 acetyl-CoA's condense to from C_6
- Lyase cleaves off acetyl-CoA to form acetoacetate
  - Some of these are reduced to 3-hydroxybutyrate
  - Some decarboxylated to acetone
    - Nonenzymatic (spontaneous)

Need for Other Lipids

- Fatty acids used for fuel or storage
  - Fuel: β-oxidation or ketogenesis
  - Storage: TAGs
    - Start with FA synthesis
    - Attach FA chains to glycerols
- Also need lipids for membranes and signaling
  - Glycerophospholipids (3 ways to make them)
    - Phosphorylate head group and add to glycerol backbone
    - Phosphorylate glycerol backbone and then add head group
    - Switch head groups

Synthesis of Triacylglycerols

- Fatty acid chains attached to phosphorylated glycerol backbone
- First activated to CoA thioesters
  - Fatty acid + CoA + ATP $\longleftrightarrow$ acyl-CoA + AMP + PP$_i$
- Acetyltransferases add fatty acid chains
- Phosphatase removes phosphoryl group

Synthesis of Phospholipids

- Head group phosphorylated
  - Phosphoryl group attacks CTP and CMP added
    - PP$_i$ released and hydrolyzed
  - OH group on C$_3$ of diacylglycerol displaces CMP
- Some produced by head group exchange
  - Serine displaces ethanolamine
- Synthesis of phosphatidylinositol
  - Phosphatidate (glycerol backbone) activated
  - Then the head group is added

Cholesterol Synthesis

Cholesterol is needed as a component of membranes, precursor for hormones, and is essential for life.

- Cholesterol is also built from acetyl units
- 1st steps similar to ketogenesis
  - Diverge after formation of HMG-CoA
- Mevalonate converted to isopentenyl pyrophosphate
  - 2 phosphoryl groups added; decarboxylated
- This isoprene derivative is the precursor of cholesterol and other isopenoids
- Six isoprene units condense to form one squalene (6 x C$_5$ = C$_{30}$)

- It takes 21 reaction to convert squalene to cholesterol

  - 30+ in total to make cholesterol

- Control step for cholesterol synthesis is the formation of mevalonate

  - One of most highly regulated enzymes known

Fates of Cholesterol

- May be incorporated into cell membranes

- May be acylated to form cholesteryl ester for storage or packaging into VLDL

- Precursor of other hormones

  - Testosterone and estrogen

- Precursor of bile acids like cholate

  - Synthesized in liver and stored in gallbladder

  - Secreted into small intestine

  - Solubilize dietary fats

  - Most recycled, but some excreted

  - Only way to dispose of cholesterol

Cholesterol Transport

- Cells can either synthesize cholesterol or take it up from LDL

  - LDL binds to cell-surface receptor and gets endocytosed

  - Lipoprotein degraded inside cell and cholesterol released

- Excess cholesterol removed from cell by HDL

  - Cholesterol moved from cytosolic leaflet of bilayer to extracellular side

  - Diffuses from there to docked HDL particle

- Since cholesterol is not really degraded and accumulation could be toxic:

  - Cholesterol shuts down its own synthesis by inhibiting HMG-CoA reductase

    - Feedback inhibition

  - Also represses transcription of gene for LDL receptor

CHAPTER 17: NITROGEN METABOLISM

Nitrogen Cycle

Nitrogen has many oxidation states: N_2, N_2O, NO, HNO_2, and NH_3. This is more than any other major element.

- Fixing nitrogen (N_2)
    - Essential for converting N into usable chemical form
    - Nonbiological processes of fixing nitrogen:
        - Lightning
        - Industry
        - Combustion
    - Biologically
        - By microbes: diazotrophs
        - Catalyzed by nitrogenase
            - Metalloprotein: Fe-S clusters; Fe-Mo cofactor
            - Sensitive to O_2: oxidizes Fe-S cofactors
        - $N_2 + 8\ H^+ + 8e^- + 16\ ATP \longleftrightarrow 2\ NH_3 + H_2 + 16\ ADP + 16\ P_i$
        - Fixing CO_2 is expensive
- NH_4^+ sources
    - Nitrates and nitrites occur naturally in water and soils
    - Reduction of NO_3^- (nitrate) to NO_2^- (nitrite) then to ammonia
        - Carried out by plants, fungi, bacteria
- Nitrification
    - Oxidation of NH_4^+ to NO_2^- (nitrite) then to NO_3^- (nitrate)
    - Some microbes can fully oxidize to nitrate – some to only nitrite
- Denitrification
    - Conversion of NO_3^- to N_2

Nitrogen Assimilation

Once we have a useful form of nitrogen (NH_3), how does it get incorporated into biological molecules?

- Glutamine synthetase (uses NTP)
  - Found in all organisms
  - Glu first activated by phosphoryl transfer
  - NH_2 exchanged for P_i
  - Activity is tightly regulated
  - Gln is carrier of amine groups
  - Glu and Gln are present in higher concentrations than other amino acids
- In bacteria and plants
  - Glutamate synthase (-ase not – tase, so no NTP needed)
  - Gln synthetase (converts Glu to Gln)
  - Glu synthase (converts α-ketoglutarate + Gln to 2 molecules of Glu):
    - α-ketoglutarate + NH_4+ + NADPH + ATP → Glu + $NADP^+$ + ADP + P_i
    - Mammals don't have Glu synthase – Glu is synthesized in other ways

Transamination Reactions

Reactions that move amino groups are termed transamination reactions.

- Transferring amine groups between molecules require the special cofactor: PLP
- Pyridoxal-5'-phosphate (PLP) carries an aldehyde group
  - There is no such group on amino acid side chains
  - Initially covalently bounded to enzyme via Lys side chain (ε-amino)
- General steps of mechanism
  - Amino acid binds first – need amine donor
    - Forms Schiff base with PLP via α-amine group
    - Breaks bond with Lys

- Steps of mechanism
  - Several chemical steps lead to hydrolysis of ketimine intermediate and release of keto acid product
    - PDX (pyridoxamine): amine group now on PLP
  - Keto acid substrate binds
    - Amine group transferred from PLP to keto acid
    - Form amino acid and regenerate PLP-Lys

Overview of Amino Acid Biosynthesis

Amino acid precursors are intermediates in glycolysis, the TCA cycle, and the pentose phosphate pathway.

- Nitrogen Metabolism in Context
  - Amino acids are synthesized from intermediates of glycolysis and the citric acid cycle
  - Nonessential amino acids can be synthesized
  - Essential amino acids must be obtained from food
- Amino Acid Biosynthesis
  - Essential amino acids
    - Unable to synthesize them
    - Sources are plants and microbes
  - Nonessential amino acids
    - Amine groups come from Glu and Gln
    - Some synthesized from essentials
      - Tyrosine from phenylalanine
      - Cysteine depends on S from Methionine

Synthesis of Nonessential Amino Acids

- Pyruvate → alanine

- Oxaloacetate → aspartate

- A-ketoglutarate → glutamate

- Glutamate → glutamine

- Aspartate → asparagine

    o Amine donor is Gln

    o Catalyzed by synthetase (costs ATP)

- Glutamate → arginine

- Glutamate → proline

    o Longer pathways

- 3-Phosphoglycerate → serine

- Serine → glycine

Synthesis of Essential Amino Acids

- Serine → cysteine → methionine (bacterial)

- Serine → cysteine (humans)

    o S atom already in amino acid

- Synthesis of aromatics begins with chorismate

    o Phe, Tyr, Trp

    o In plants and bacteria

- Synthesis of Tyrosine in humans

    o Depends on Phe

- Synthesis of Histidine

    o Other 19 amino acids synthesized by pathways of carbohydrate metabolism

    o Histidine synthesized from ATP + Glu + Gln + PRPP

Synthesis of Neurotransmitters

Amino acid-based neurotransmitters:

- Γ-aminobutyric acid (GABA): Glu derivative

    o Regulation of muscle tone

- Tyrosine → catecholamines

    o Dopamine (reward behavior)

        ▪ Parkinson's disease

    o Norepinephrine, epinephrine

- Tryptophan → serotonin

    o Linked to depression

- Tryptophan → melatonin

    o Correlation with circadian rhythms

- Amino acid-based neurotransmitters

    o Arginine → •NO (g)

        ▪ Nitric oxide radical

        ▪ Can't be stored

        ▪ Diffuses into cell

        ▪ Breaks down on its own

        ▪ Produced only when and where needed

        ▪ Induces blood vessel dilation

            • Nitroglycerin → to relieve angina

        ▪ Kills pathogens at high concentration

Amino Acid Catabolism

- Classification of amino acids
    - Glucogenic: give rise to gluconeogenic precursors like TCA intermediates
    - Ketogenic: give rise to acetyl-CoA that can be used for ketogenesis or FA synthesis
- Glucogenic Amino Acids
    - Transaminations
        - Alanine → pyruvate
        - Aspartate → oxaloacetate
        - Glutamate → α-ketoglutarate
    - Deamidation + transamination
        - Asn → ASP → oxaloacetate
    - Deamidation
        - Serine → pyruvate
        - Cysteine → pyruvate

Branched Chain Amino Acids

- Val (glucogenic), Leu (ketogenic), Ille (Both)
- More complicated pathways
- Threonine → acetyl-CoA + glycine
    - Glycine → serine → pyruvate
- Another pathway for glycine catabolism
    - Glycine cleavage system
        - Deamidation
- Aromatics
    - Tyr, Trp, Phe
    - Example of Phe
        - Produces acetoacetate (ketogenic) and fumarate (glucogenic)

Purine Synthesis

Nucleotides can be recycled from nucleic acids and cofactors that are broken down, they can also be synthesized de novo from several amino acids. These processes work so well that there are no specific dietary requirements for purines and pyrimidines.

- Purines = AMP & GMP

- Base builds onto sugar-phosphate

  - PRPP made first

- 10 steps

  - Require: Gln, Gly, Asp, HCO_3^-, and H-C=O from tetrahydrofolate

  - Product: IMP

  - Base: hypoxanthine

- For AMP: need $-NH_2$ from Asp

- For GMP: need $-NH_2$ from Gln

- Kinases add P groups

- Regulation

  - High [GTP] ↑ AMP

  - High [ATP] ↑ GMP

  - Ensures balance of purines

  - 1st step (PRPP)

    - Feedback inhibited by ADP/GDP

Pyrimidine Synthesis

- Base built first, then attached to PRPP

- Requires: Gln, Asp, HCO_3^-

- Product is UMP

- Kinases convert UMP to UDP and UTP

- CTP synthetase makes CTP from UTP

- Regulation
    - Feedback inhibition by UMP, UDP, UTP
    - ATP activates 1st step in synthesis
        - Balances levels pf purines and pyrimidines

Synthesis of Deoxyribonucleotides

- Nucleotide triphosphates 1st converted to diphosphates
    - NTP → NDP
- Ribonucleotide reductase converts 2' –OH to –H
    - NDP → dNDP
- dNDP then converted to dNTP
- This produces all 4 deoxynucleotides
    - dATP, dGTP, dCTP, dUTP
- Making dTTP
    - dUTP rapidly converted to dTTP
        - dUTP hydrolyzed to dUMP
        - Methyl group added
        - dTMP converted to dTTP

Nucleotide Catabolism

The products of nucleotide breakdown are ribose groups, purine bases, and pyrimidine bases.

- **Purine Catabolism**
    - Purine bases are converted to uric acid
    - Uric acid is excreted as waste
        - Excess of uric acid in kidneys forms crystals (kidney stones)
    - Uric acid can also get deposited in joints to cause gout
    - In other organisms
        - Uric acid is converted to urea or ammonia

- **Pyrimidine Catabolism**
    - Pyrimidine ring is first opened
    - Then deaminated and reduced
    - Broken down to β-alanine or β-aminoisobutyrate
        - Breakdown products feed into other metabolic pathways
        - Not excreted as waste

Urea Cycle

The Urea Cycle is way to get rid of amine groups resulting from amino acid catabolism.

- High concentrations of NH_4^+ is toxic
    - Causes alkalosis
    - Can enter the brain and combine with α-ketoglutarate to form Glu
        - This is bad because it depletes TCA intermediates
    - Most nitrogen is excreted as urea

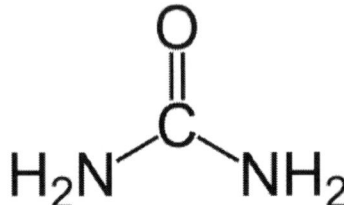

Structure of Urea:

- Glutamate dehydrogenase reaction is the major way to feed amino groups into urea cycles
    - Glutamate dehydrogenase can use either NAD^+ or $NADP^+$ as a cofactor
    - Releases NH_4^+ group
- First step in incorporating the NH_4^+ that was released into the urea molecule is the carbamoyl phosphate synthetase reaction (-tase so needs NTP)
    - Forms a carboxyphosphate intermediate
    - NH_3 may come from Glu dehydrogenase reaction or another reaction
    - Costs 2 ATP per carbomoyl phosphate

- Energy Cost

  o 4 ATP to make 1 urea molecule

- Regulation of Cycle

  o Regulated mostly by the activity of carbamoyl phosphate synthetase

    ▪ Activated by N-acetylglutamate

      • Signals that there is plenty of $-NH_2$ donors to feed into the cycle

    ▪ Production of NAG promoted by high cellular levels of Glu (amine carrier) and acetyl-CoA

CHAPTER 18: REGULATION OF MAMMALIAN FUEL METABOLISM

Regulating Metabolism

The human body needs to convert the energy that is taken in from food in order to perform various functions. Food intake is variable (we don't always eat the same thing or the same amount) and different organs and tissues must communicate in order to operate as a unit. This is the reason there is a need for regulation on a larger-scale.

- Compartmentation: keep processes spatially separate

- Hormonal control: use outside signal to stimulate change inside cell

- Cellular Compartmentation

    - Anabolic and catabolic pathways regulated

        - Counterproductive to have both operating at the same time

    - Opposing processes may be spatially separated into different cellular compartments

    - This necessitates transport of metabolites across membranes

Liver Functions

Liver in a Fed State

- Glucose is stored as glycogen

- Excess glucose and amino acids are catabolized to acetyl-CoA

    - Acetyl-CoA is used to synthesize fatty acids

    - Converted to triacylglycerols (TAGs) and sent to other tissues

Liver in a Fasted State

- Breaks down glycogen to glucose and then sends it to other tissues

- TAGs broken down to acetyl-CoA and it can then be used to make ketone bodies

- Amino acids can be broken down and converted to glucose or ketone bodies

Liver in a Fed or Fasted State

- Processes lactate and alanine produced by muscles

- Disposes of amine groups through urea cycle

Kidney Tie-In

- Kidneys: eliminate waste; maintain pH balance

  - Deaminate glutamine to form α-ketoglutarate

  - Used in gluconeogenesis (only liver and kidney can do this)

Muscle Functions

Muscles in a Fed State

- Take up glucose and store it as glycogen

  - Only a limited amount of glycogen can be stored in muscles

Muscles in a Starved State

- Protein broken down to amino acids

  - Used to generate glucose by the liver

Muscles in an Active state

- Glycogen broken down for glycolysis

  - Produces ATP

  - Lactate and alanine are produced & exported

    - Processed by liver

- Fatty acids and/or ketone bodies broken down to acetyl-CoA

- Heart muscle uses fatty acids as primary fuel

  - Rich in mitochondria (site of β-oxidation)

Adipocytes

Adiopocytes in a Fed State

- Take up glucose and convert it to glycerol

- Fatty acids taken up and combined with glycerol to make TAGs

  - TAGs are stored as fat globules

Adiopocytes in a Fasted State

- Fatty acids mobilized from TAGs

- Released into circulation

Metabolite Movement

A metabolite is a substance formed in or necessary for metabolism.

- Circulatory system delivers metabolites from and to organs

    o Glucose produced by liver travels to other tissues

    o Amino acids travel to liver or kidney for amine disposal

- Necessary because not all organs and tissues have the means to utilize every metabolic pathway

    o Substrates may come from another organ or tissue

    o End product may need to be delivered to a different organ or tissue

Cori Cycle

The Cori Cycle is also known as the Lactic Acid Cycle.

- Muscle breaks down glycogen during periods of high activity

    o Glycolysis generates ATP needed for muscle contraction

    o Rapid catabolism generates NADH faster than it can be oxidized by the mitochondria

    o Lactate is generated to reoxidize NADH

        ▪ Travels to liver to participate in gluconeogenesis

        ▪ ATP needed comes from oxidation of fatty acids

    o Glucose generated in liver travels back to muscle

- Net effect: transfer of energy from liver to muscle

Glucose-Alanine Cycle

This cycle is another link between the liver and muscles.

- Muscle protein breaks down (due to vigorous exercise)

    o Amino acids deaminated to produce TCA intermediates

- - - o Pyruvate transaminated to produce **alanine**
- Alanine travels to liver
    - o Alanine deaminated to feed into urea cycle
    - o Pyruvate used in gluconeogenesis
- Glucose travels back to muscle
- Net effect: transport of nitrogen from muscles to liver

Hormone Signaling

Activities of organs that store and release fuel regulated and coordinated by hormones.

- Produced in one tissue – influence functions of other tissues
    - o Most important in fuel metabolism
        - ▪ Insulin
        - ▪ Glucagon/catecholamines (epinephrine/norepinephrine)
    - o Many others involved in appetite control, fuel allocation, body weight
    - o Hormones bind to receptor to exert some cellular response
- Insulin Release
    - o Insulin is released in response to rise in blood glucose levels
        - ▪ Normal [glucose]: 3.6 to 5.8 mM; Fed state: 8 mM
        - ▪ Insulin is synthesized in β cells in the pancreas
    - o Insulin released in response to rise in blood glucose levels
        - ▪ Trigger mechanism for insulin release is not well understood
            - • β cells do not have glucose receptor (do have a transporter)
            - • Trigger driven by glucose metabolism
        - ▪ One glucose sensor: glucokinase (in liver and pancreatic β cells)
            - • Isozyme of hexokinase
            - • Glucose + ATP → G-6-P + ADP

Glucose Sensor

- Hexokinase: K_M (glucose) < 0.1 mM

  - Enzymes saturated with substrate at physiological levels

- Glucokinase: K_M (glucose) 5-10 mM

  - Never fully saturated (Not working at V_{max})

    - Greatest sensitivity to substrate concentrations

  - Substrate saturation curve is sigmoidal

    - Allostery – for a single polypeptide

  - May be due to substrate-induced conformational changes

    - Enzyme in R (high affinity) state at end of catalytic cycle

    - If [Glc] high, readily binds substrate and reaction velocity high

    - If [Glc] low, reverts to T (low affinity) state before substrate binds

Comparison Graph of Hexokinase and Glucokinase Activities:

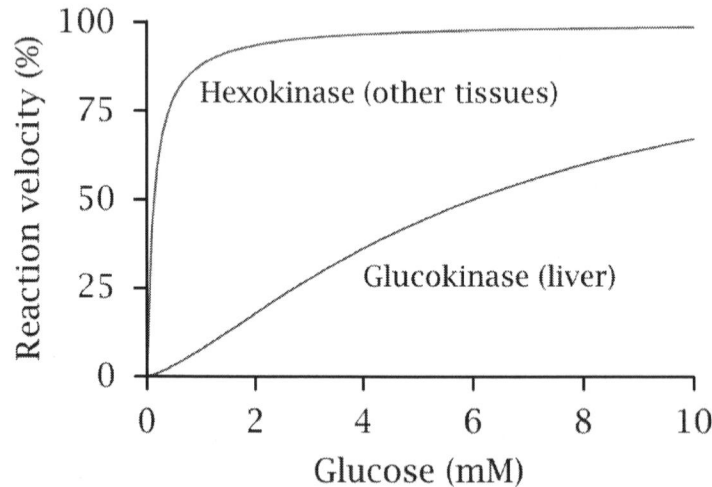

Insulin Effects

- Not all cells have insulin receptors

  - Response is tissue specific

- 1st effect: glucose uptake

  - GLUT4 glucose transporters localized to membranes of intracellular vesicles

- - Binding of insulin to receptor stimulates translocation of GLUT4 glucose transporters in muscle and adipose tissue
    - Stimulates fusion of vesicles with plasma membrane
  - Opens the cellular doors to let glucose in (↑ uptake)
    - Regular exercise increases # of GLUT4 transporters
- Insulin Signals
  - Insulin signals that glucose is available in abundance
    - Stimulate usage and storage
    - Inhibit mobilization of stored fuel
  - Stimulates glycogen storage
    - Activate glycogen synthase
  - Inhibit glycogen breakdown
    - Inhibit glycogen phosphorylase
  - Both accomplished by dephosphorylation
- Global Effects of Insulin
  - Global effect is to burn fuel and store excess for later
  - Some of glucose used by liver to synthesize FAs & TAGs
    - Sent to tissues as VLDL/LDL
  - Insulin stimulates lipases to release FAs from lipoproteins to be taken up by adipocytes and stored as TAGs

Effects of Glucagon

- Glucagon is small (29 residue) peptide hormone
  - Synthesized and released by pancreatic α cells
- Produced in response to drop in blood glucose levels
- Stimulates production of glucose by liver
- Stimulates lipolysis in adipose tissue
  - Mobilize stored TAGs
- Muscle cells respond to a different hormone (Epinephrine)

Other Regulatory Hormones

Leptin

- Functions as an appetite suppressor

- Accumulation of fat elevates levels of leptin produced by adipocytes

- Leptin travels to brain to signal satiety

- Decreases acetyl-CoA carboxylase activity

  - Decreases Malonyl-CoA; decreases FA synthesis; increases β-oxidation

Adiponectin

- Activates AMP-dependent protein kinase

  - Turns on ATP-production & turns off ATP-consuming pathways

  - Inhibited by ATP

- Increased catabolism of glucose and fatty acids

- Increased sensitivity to insulin

Resistin

- Blocks activity of insulin

Obesity

- Significant associated health risks

- No single cause

- Lack of leptin gene in mice or humans causes severe obesity

- May relate to set-point weight

- Obesity may also result from leptin resistance

  - High set-point

Types of Fat

- Subcutaneous fat (beneath the skin)

- Visceral fat (surrounding the abdominal organs)

  - Large amounts correlate with cardiovascular disease and type II diabetes

- Brown fat (high mitochondrial content)

  o More like muscle than fat

  o May relate to ability to burn off excess fuel

Diabetes Mellitus

- Hyperglycemia: high blood glucose

  o High levels in urine

  o Since muscle and adipose tissue can't take up glucose – tends to enter other tissues

    - Glucose there converted to sorbitol

    - Affects osmotic balance: alters kidney function and triggers protein precipitation in other tissues

- Also a disorder of fat metabolism

  o Remember: insulin stimulates synthesis of TAGs

  o Metabolize fatty acids rather than carbs

    - Production of ketone bodies (ketoacidosis)

- Type I (juvenile onset)

  o Autoimmune disease

  o Immune system destroys pancreatic β cells

    - Unable to make insulin

  o First treatments involve administration of pancreatic extract

  o Administer exogenous insulin

  o Insulin-dependent diabetes

- Type II (adult onset)

  o Due to insulin resistance

  o Cells fail to take up glucose

  o Liver responds by making more – increasing blood glucose levels more

- Variety of drugs used to compensate for insulin resistance
    - Suppress gluconeogenic enzymes
    - Activate AMP-dependent kinase

Metabolic Syndrome

Set of symptoms including obesity and insulin resistance that appear to be related.

- Often manifests as type II diabetes – with atherosclerosis and hypertension
- Higher risk for cancer
- Higher proportion of visceral fat

CHAPTER 19: GENES TO PROTEINS

Nucleotides and Nucleic Acids

Nucleotides (Monomers of Nucleic Acids)

- 3 components of nucleotides:

    o 5-carbon sugar (pentose)

    o Nitrogen containing base ring

    o PO_3 group (phosphate group)

        ▪ Attached to the #5 carbon of the sugar

- Pyrimidine – nitrogenous base with a characteristic six-membered ring consisting of carbon and nitrogen atoms

    o Cytosine (C), Thymine (T), and Uracil (U) are pyrimidines

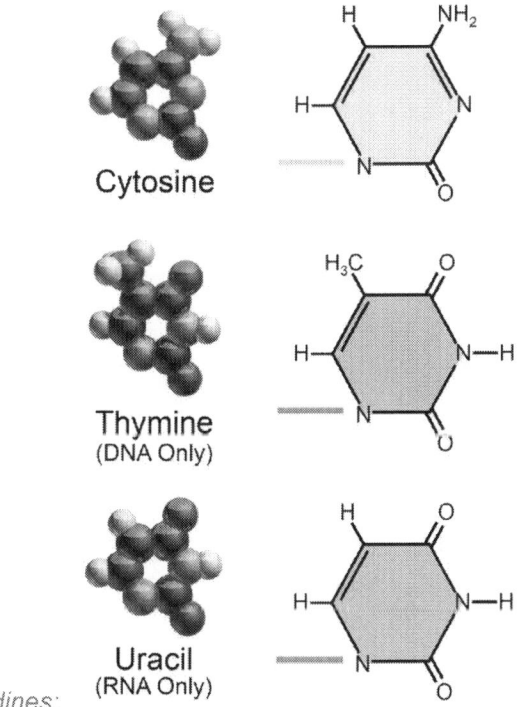

Cytosine

Thymine
(DNA Only)

Uracil
(RNA Only)

Pyrimidines:

- Purine – nitrogenous base with a characteristic five-membered ring fused to a six-membered ring

    o Adenine (A), and Guanine (G) are purines

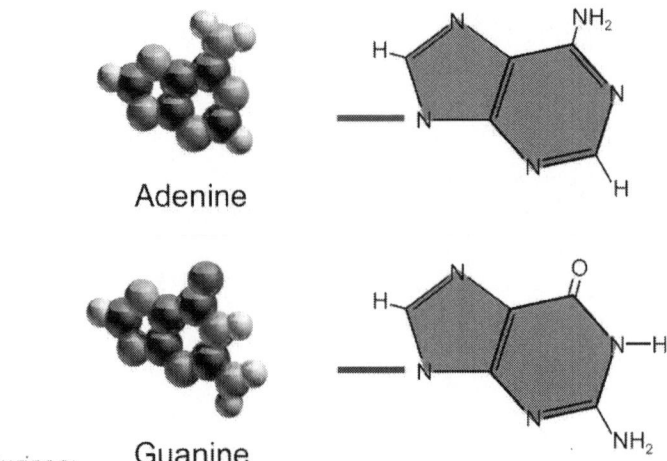

Adenine

Guanine

Purines:

- Phosphodiester bonds covalently link nucleotides together

  - Phosphate connects the 5' carbon of one nucleotide to the 3' carbon of another nucleotide

    - Gives strands directionality

    - In a strand, all sugars are oriented in the same direction

- Phosphates and sugar form the backbone of a strand

  - Bases project from the backbone

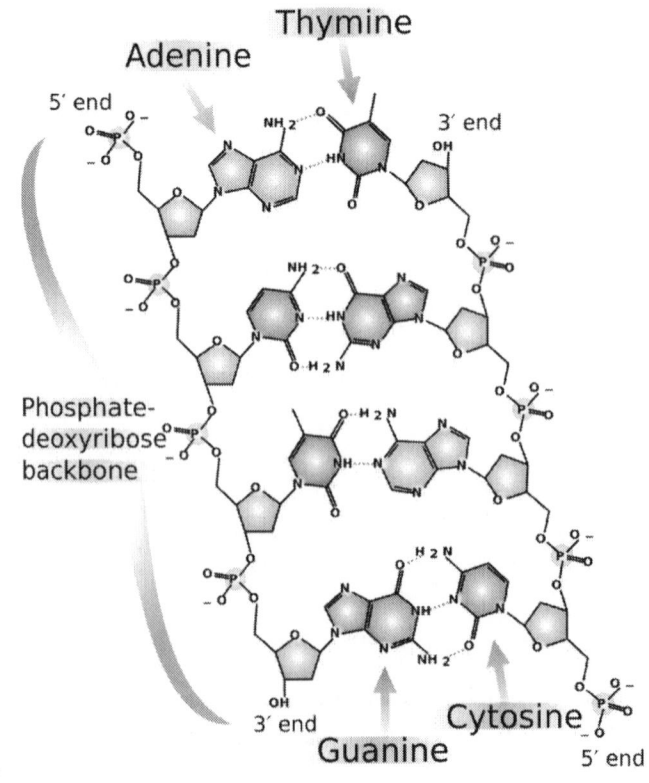

DNA Structure:

DNA (Deoxyribonucleic Acid)

- Deoxyribose as the pentose sugar

- Two nucleotide chains form a double helix

- Contains the nucleotides:

    o Thymine (T), Adenine (A), Cytosine (C), and Guanine (G)

        ▪ A forms 2 hydrogen bonds with T

        ▪ G forms 3 hydrogen bonds with C

- Contain genes that are instructions for protein synthesis

- Sugar-phosphate backbones are solvent-exposed

    o Hydrophilic

- Base pairs are perpendicular to backbone

    o Bases are mostly hydrophobic

    o Need to minimize solvent exposure

    o Base pairs stack closer together

    o Movement of bases together tilts the backbone by 30

    o Creates an uneven twist

        ▪ Major groove vs. minor groove

            • These are spaces between the backbones

        ▪ Strands not directly opposite each other

    o Stability of DNA due to double-stranded helix

        ▪ Has most to do with base-stacking interactions and hydrophobicity

            • Van der Waals interactions

            • Cumulative effect

        ▪ H-bonding of bases also a factor

        ▪ Ionic interactions

            • Cations (e.g. Mg^{2+}, Na^+, K^+)

- DNA Base Stacking
  - Pi orbitals overlap in benzene
    - Alternating single and double bonds
    - Pi conjugation
    - Very stable – electrons (e^-) have more freedom of movement
  - Bases in DNA also have alternating single and double bonds
    - They also have pi orbitals
    - Pi orbitals of adjacent base pairs can overlap
    - Greater freedom of movement for e^-

RNA (Ribonucleic Acid)

- Single stranded nucleic acid
- Ribose is the pentose sugar
- Contains same nucleotides as DNA except that Adenine (A) is replaced with Uracil (U)
- Functions in the synthesis of proteins
  - Messenger RNA (mRNA) carries encoded genetic message to the cytoplasm from the nucleus

Terminology

- DNA size expressed as "bp" (base pairs) or as "kb" (kilobase pairs)
- Oligonucleotide or oligo = short single stranded DNA molecule
- DNA synthesized by polymerases
- DNA digested by nucleases
  - Exonuclease digests from the end ("exo" = "outside")
  - Endonuclease cuts at specific site ("endo" = "within")

Melting (Denaturing) DNA

Melting DNA (or denaturing DNA) refers to separating the 2 strands by breaking H-bonds and base stacking interactions.

- Thermal energy needs to be greater than the intermolecular bond energy to separate strand

- A:T separates more readily than G:C

  - Remember – this is due to base stacking differences

  - (GC)-(GC): -61 kJ/mol

  - (AT)-(AT): -27.5 kJ/mol

- The amount of GC content is directly related to melting temperature

- Length of DNA molecule also important

- T_m is the temperature at which ½ of the DNA is melted

Renaturing DNA

- Temperature can be lowered to re-anneal DNA strands

  - Re-form H-bonds and base stacking

- Rapid cooling results in mismatched base pairing

- Annealing time depends on the length of the DNA molecule

The Central Dogma

The central dogma describes the flow of genetic information in cells from DNA to RNA and to protein.

- DNA is replicated with each division cycle

  - Replication is semi-conservative

    - One parent strand is conserved in the new DNA molecule

- DNA is transcribed into mRNA

  - Transcribe one form of nucleic acid into another form

- mRNA is translated into protein

  - Language of nucleic acid translated into language of amino acids

- Notable exceptions to the dogma are viruses

Translation: Protein Synthesis

- Translation occurs on the ribosome, which contains rRNA and many other proteins
- In translation, tRNA carries amino acids to the ribosome and binds to its complement in the mRNA template
- Amino acids are dictated by the genetic code

Genetic Code

The genetic code is a set of rules by which information encoded within genetic material is translated into proteins by living cells.

Seond letter

First letter	U	C	A	G	Third letter
U	UUU Phe	UCU Ser	UAU Tyr	UGU Cys	U
	UUC Phe	UCC Ser	UAC Tyr	UGC Cys	C
	UUA Leu	UCA Ser	UAA Stop	UGA Stop	A
	UUG Leu	UCG Ser	UAG Stop	UGG Trp	G
C	CUU Leu	CCU Pro	CAU His	CGU Arg	U
	CUC Leu	CCC Pro	CAC His	CGC Arg	C
	CUA Leu	CCA Pro	CAA Gin	CGA Arg	A
	CUG Leu	CCG Pro	CAG Gin	CGG Arg	G
A	AUU Ile	ACU Thr	AAU Asn	AGU Ser	U
	AUC Ile	ACC Thr	AAC Asn	AGC Ser	C
	AUA Ile	ACA Thr	AAA Lys	AGA Arg	A
	AUG Met	ACG Thr	AAG Lys	AGG Arg	G
G	GUU Val	GCU Ala	GAU Asp	GGU Gly	U
	GUC Val	GCC Ala	GAC Asp	GGC Gly	C
	GUA Val	GCA Ala	GAA Glu	GGA Gly	A
	GUG Val	GCG Ala	GAG Glu	GGG Gly	G

The Genetic Code:

- Triplet codons
  - 64 codons
  - 20 common amino acids
  - 3 stop codons
- Code is degenerate
  - More than one codon for most amino acids
  - Third base "wobble"
  - Except methionine: it is always the start codon
- Code is universal

Transcription

- RNA synthesis proceeds by complementary base pairing with one of the DNA strands

- This strand acts as the template to determine which ribonucleotide gets added and in what order

- The DNA strand that guides its synthesis is therefore the "noncoding" or template strand

- The mRNA synthesized contains the genetic code

- The complementary strand of DNA is therefore the coding strand

- The mRNA product is therefore an RNA copy of the coding strand of the DNA (U in place of T)

Coding and Noncoding DNA

- Prokaryotes

  - All but a few percent of the DNA are genes for proteins and RNA

- Eukaryotes

  - Proportion of noncoding DNA increases with organismal complexity

  - >98% of human is noncoding (not expressed as protein)

  - Up to 80% transcribed to RNA (but not translated to protein)

- Transposable elements

  - "Jumping genes"

  - DNA segments copied and pasted multiple times

  - Repetitive DNA sequences

Identifying Genes

The region of the nucleotide sequence from the start codon to the stop codon is the open reading frame.

- Looking for an Open Reading Frame (ORF)
    - Computer scans the DNA sequence
        - Looks for universal start codon (ATG/AUG)
        - Also looks for translational stop codon (1 of 3)
        - Looks for longest ORF
    - This is a rough method that often over-estimates the gene number
- Sequence comparison with known genes
    - Genome to genome comparison
    - Possible because of universal nature of genetic code
    - Tends to underestimate number of genes
        - Requires that we already know the sequence in at least one organism
- Benefits of Genomics
    - Identify human genes with homologues in model organism
        - Mutate gene in a model organism to determine its function then extrapolate to how it would affect humans
    - Can be used to identify disease markers

CHAPTER 20: DNA REPLICATION AND REPAIR

DNA in Context

- Complete cell DNA sequence = genome
  - Genome of bacteria is usually circular
    - Can be linear
- Bacterial genome = 0.6 - 9.4 million bp
- Human genome = 4 billion bp
  - 1000x larger than *E. coli*
  - But only 8X the genes: 30,000 (us) vs. 4,000 (*E. coli*).
  - *E. coli* genes use less of the DNA sequence
    - About 1000 bases in a typical bacterial gene
    - 3000 to 2.4 million in a human gene

DNA Supercoiling

- Twist and Writhe
  - Twist (T)
    - Number of helical turns of one strand around the other
  - Writhe (W)
    - # of times double helix crosses over itself
    - Supercoiling
  - Linking number (L)
    - $L = T + W$
    - Represents the amount of tension in the molecule
  - Separation of DNA strands ↑ supercoiling in advance of separation
    - Important consideration in DNA replication

- Supercoiling compacts DNA
  - Unsupercoiled DNA = 1 twist for ~10 bases
  - Positive supercoils
    - Supercoils in left-handed direction
    - Winding more frequent
      - Overwinding
      - Harder to unwind
  - Negative supercoils
    - Supercoils in right-handed direction
    - Winding less frequent
      - Underwinding
  - Supercoils twist DNA

Topoisomerases

Type I Topoisomerases

- Relieve torsional stress caused by supercoils
  - Alter Twist
- Make single-strand break
- Work on both positive and negative supercoiling
- Energy released

Type II Topoisomerases

- Make double-strand break
- Requires energy
  - ATP hydrolysis
- Can relieve both positive and negative supercoiling
  - Alter Writhe
- DNA gyrase (prokaryotes)
  - Introduces more negative supercoiling

- Eukaryotes maintain negative supercoiling by wrapping DNA around nucleosomes

Archaeal Topoisomerases

- Increases linking number
  - Results in positive supercoils
  - Harder to unwind
  - More resistant to stress

<u>Models of Replication</u>

- Replication Fork

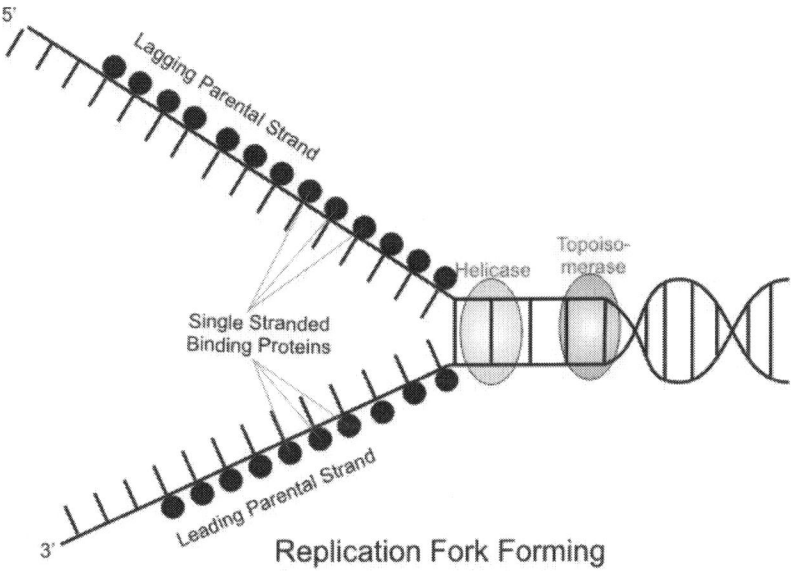

Replication Fork Forming

  - Where parental strands separate and new strands are synthesized
  - Begins at *Ori* site (origin of replication) in bacteria
  - Begins at numerous sites in eukaryotic cells
- Replication Models
  - Locomotive model of replication
    - Replication machinery moves along DNA like train on tracks
  - Factory model of Replication
    - Replication machinery is stationary and feeds DNA through
    - This is the true model

Obstacles to DNA Polymerase

- DNA polymerase synthesizes from the 5' to 3' end

  - Reads 3' to 5'

  - 3' OH of the last nucleotide attacks the phosphate of the incoming nucleotide

  - Breaks phosphoanhydride bond of dNTP; forms phosphodiester bond in DNA

    - $\Delta G \approx 0$

    - But it is irreversible due to the hydrolysis of PPi

  - DNA polymerase needs to have a 3'-OH to add onto

  - Has proofreading ability

- Obstacles to DNA Polymerases

  - Can only extend existing chain

    - Needs 3'-OH to build on

    - The solution: use RNA polymerase (DNA primase) to make RNA primer

      - Primer can be removed later

  - DNA is antiparallel

    - No problem on strand that runs 3' to 5'

    - Problem encountered with strand that is 5' to 3'

    - The solution:

      - Synthesize leading strand continuously

      - Synthesize lagging strand discontinuously

        - Form Okazaki fragments (100-200 nucleotide in humans)

DNA Replication – Overview of Steps

- Topoisomerase relaxes supercoiling by nicking one strand
- Helicase unwinds the helix (separates the 2 strands)
    - Need proteins to protect single stranded DNA
        - SSB: single-strand binding protein
- RNA primers are synthesized
- DNA replicated by DNA polymerase
- RNA primers removed & replaced with DNA
- Gaps sealed by ligase
- Supercoiling added

DNA Polymerase

Structure

- 2 Mg^{2+} ions coordinated by Asp residues and phosphate groups of substrate NTP
- 1 Mg^{2+} ion interacts with 3' O atom to make it more nucleophilic
- 2' deoxy lies in hydrophobic pocket
    - Won't bind RNA (has a 2'-OH)

Mechanism

- DNA polymerases are processive
    - Catalyze consecutive reactions without releasing their substrate
        - Sliding clamp helps hold the polymerase in place

Types of polymerases

- E. coli has 5
- Eukaryotes have ~12
    - Perform different roles
        - Synthesize leading or lagging strands

Proofreading

- 3' to 5' exonuclease activity acts as proofreader

  o Clips off wrong nucleotide if wrong one binds

  o Limits error rate to 1 in 10^6

  o Also a form of DNA repair mechanisms

DNA Ligase and the Lagging Strand

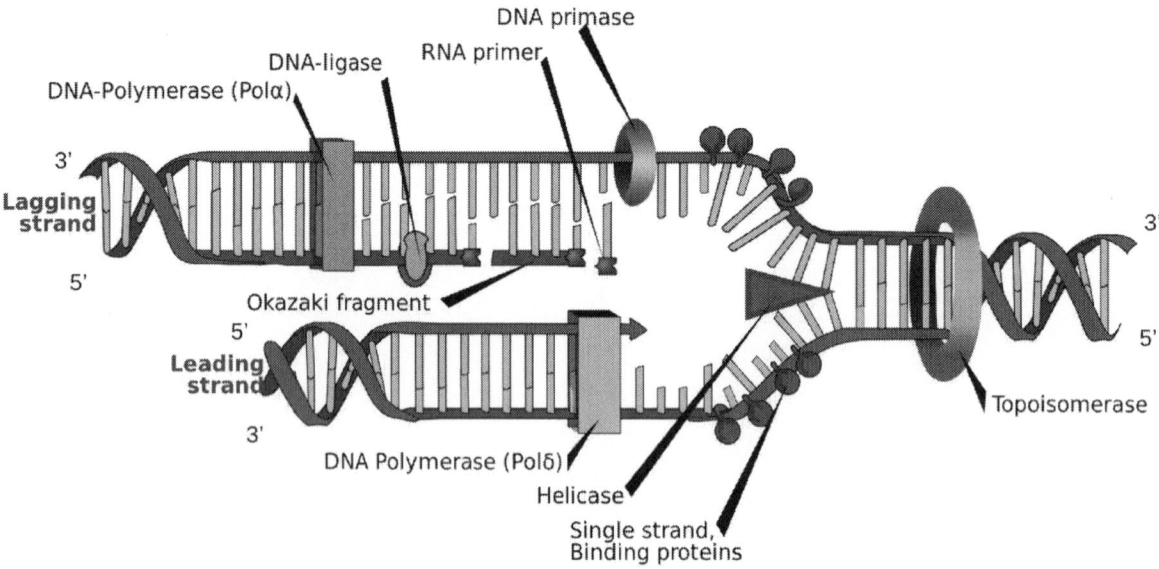

- Okazaki fragments need to be joined to make a continuous DNA strand

- RNase H is a 5' to 3' exonuclease that removes the RNA primer

  o May continue to work until DNA polymerase catches up to it (RNase H is slower)

  o Some of the older Okazaki fragment may also be removed

  o Ensures fidelity of replication

    ▪ Higher error rate in the 1st few nucleotides added

- If DNA polymerase gets there before RNase H, it can displace RNA primer to create single-stranded flap

  o Flap endonuclease cleaves off the flap (recognizes junction between RNA and DNA)

- Joining Okazaki Fragments
    - DNA Ligase joins discontinuous fragments
    - AMP used to activate 5' phosphate on DNA strand
    - Transferred from
        - ATP in eukaryotes
        - NAD in prokaryotes

Telomeres

A telomere is a region of repetitive nucleotide sequences at the end of chromosomes, which protects the ends of chromosomes from deterioration during DNA replication.

- Bacterial chromosomes are circular
    - Replication terminates where two replication forks meet
    - No need for telomeres
- Eukaryotic chromosomes are linear
    - There is no problem at 5' ends of the template strand
    - There is problem at 3' ends of template strand
        - There is no way to replace the RNA primer with DNA
        - So each round of DNA replication leads to chromosome shortening
    - Creates a need for telomeres
- Telomerase
    - Adds sequence of 6 nucleotides to 3' end of DNA strand
        - Synthesis proceeds in 5' to 3' direction
        - Repeats of TTAGGG sequence (2-10 kb long)
    - Adds sequence of 6 nucleotides to 3' end of DNA strand
        - Uses RNA molecule (~451 nucleotides) as template
        - RNA is part of the structure of telomerase

- Reverse transcriptase activity (RNA → DNA)

- When 3' end is extended, DNA polymerase can then synthesize the complement

- When replication ends there still is a shortened DNA segment, but the telomere is shortened and not the genome

DNA Damage

- Mutation is the alteration of a cell's DNA

    o Single-cell parent passes it to daughter cells

    o Multicellular organisms pass mutations to offspring only if the mutation is in reproductive cells

        - Otherwise it affects only the progeny of original cells

- Mutation can be beneficial or harmful

- Badly damaged DNA mostly results in apoptosis (programmed cell death)

Types of DNA Damage

- Point mutation (substitution) – single nucleotide substitution

    o Two types:

        - Transition: purine changed to another purine or pyrimidine changed to another pyrimidine

        - Transversion: purine changed to pyrimidine or vice versa

- Deletion – deletion of at least 1 nucleotide

- Insertion – insertion of at least 1 nucleotide

- Mistakes by DNA polymerase

    o Mis-matched bases

- Damage by ROS (Reactive Oxygen Species) like $\cdot O_2-$ or H_2O_2

    o By-products of oxidative metabolism

    o Example: G oxidized to oxoG

        - Can base pair with either C or A

- Spontaneous depurination
    - Glycosidic bond connecting base to sugar is broken
    - Results in abasic site
    - Occurs ~18,000X/day
- Spontaneous deamination
    - Removal of an amine group
    - Original G:C base pair can become T:A
    - Occurs ~500 times per day

Mutagens Cause Mutations

- Mutation Sources
    - Electromagnetic radiation
    - X-rays, gamma rays
    - UV light
        - Cause pyrimidine dimers

DNA Packaging and Modification

DNA packaging is an important process in order for a cell to be able to accommodate the large amount of DNA that it contains.

DNA Packaging

- Eukaryotic DNA is highly condensed
    - Heterochromatin (HC)
        - Densely packed; stains intensely
        - No transcription
    - Euchromatin (EC)
        - Less condensed; stains less well
        - Transcribed at a higher rate

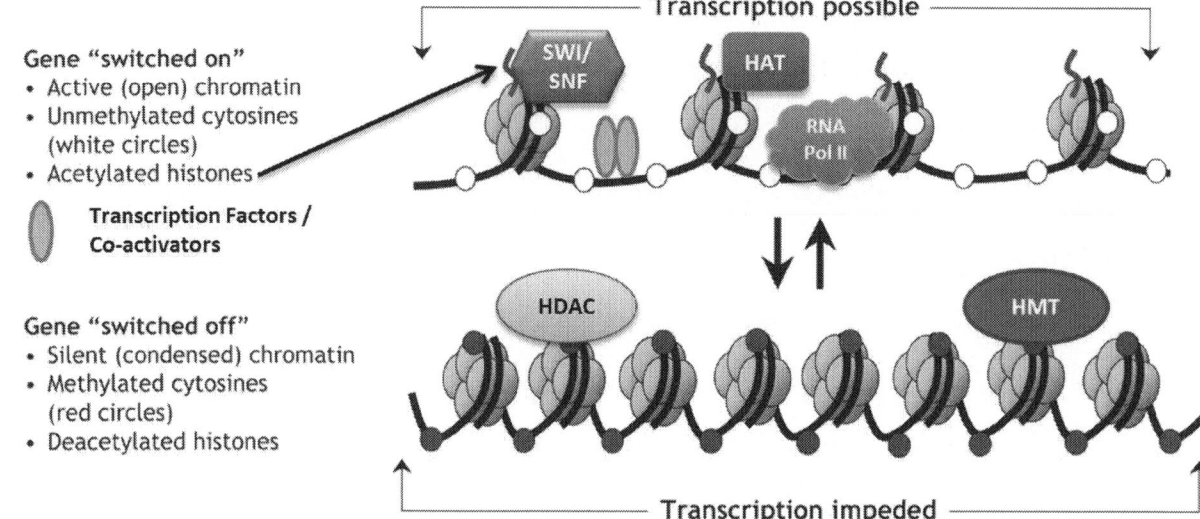

Gene "switched on"
- Active (open) chromatin
- Unmethylated cytosines (white circles)
- Acetylated histones

Transcription Factors / Co-activators

Gene "switched off"
- Silent (condensed) chromatin
- Methylated cytosines (red circles)
- Deacetylated histones

Transcription possible

Transcription impeded

- Levels of Chromatin Structure

  o Fibers loop to form packed chromosome

  o Nucleosomes aggregate to form 30-nm fiber

  o DNA wraps around histones to form a nucleosome

Modifications to Chromatin

- Histone modification

  o Histone pairs associate (1 long helix + 2 short helices)

  o Tails of histones are flexible and charged – extend out from core

  o Modifying binding of histones to DNA impacts the degree to which transcription is possible

    ▪ Methylation ($-CH_3$)

    ▪ Acetylation ($-COCH_3$)

      • Catalyzed by HATs (histone acetyl transferases)

      • Added to Lys residues from acetyl-CoA

      • Reversed by histone deacetylase

    ▪ Phosphorylation ($-OPO_3^{2-}$)

    ▪ Adding or removing chemical groups is often the switch from transcriptionally silent to active chromatin and vice versa

DNA Modification

- C residues next to G residues are methylated (mammals)

    o ~80% of CG sequences modified in this way

- Methyl group projects out into major groove – harder for DNA binding proteins to bind sequence

- CpG islands (C-phosphate-G)

    o Clusters of CpG sequence located near gene promoters

    o These are typically unmethylated

    o More readily accessible for transcription

- Methylation is way to silence DNA containing no genes

Dideoxy DNA Sequencing

- The reaction contains:

    o Template DNA - What you are interested in sequencing

    o Primer DNA - DNA polymerase needs something to build on

    o DNA polymerase – to synthesize new DNA

    o dNTPs (4) - Substrates

    o ddNTPs (4, with 4 different fluorescent labels) - To stop the reaction

- Why the reactions stop

    o No 3' –OH on ddNTP for polymerase to build onto

    o Different fluorescent labels on ddNTPs allows for the determination of which is which

    o A light detector reads and identifies the fluorescent tag

    o Computer attached to sequencer provides print out of the sequence

PCR

Polymerase chain reaction (PCR) is a method used to amplify a single copy or a few copies of a DNA sequence to produce a large number of copies.

- Involves 3 cycles

    o Denaturation of DNA (melting)

    o Primer annealing

    o Primer extension

- These cycles are repeated ~25-35 times

- Primer in one cycle becomes the template in the next

- Solution is repeatedly heated and cooled to perform each cycle

    o Requires a thermostable DNA polymerase

        ▪ Taq (Thermus aquaticus) and Pfu (Pyrococcus furiosus) are commonly used

Restriction Enzymes

Restriction enzymes are also known as restriction endonucleases which digest (cut) double stranded DNA at specific recognition sites.

- Recognition site is palindrome

    o Word example: "Racecar" – spelled same forward and backward

- Recognition sites are different sizes and unique sites

- Cuts between same nucleotides on each strand

- Always binds to double stranded DNA

- Will cut any piece of DNA that contains the recognition site

- Most restriction enzymes make staggered cuts

    o Leave "sticky ends" – an overhang of single stranded DNA

DNA Cloning

Recombinant DNA technology

- 2 separate DNA molecules are "recombined" to form a new molecule

    o A vector or plasmid is the carrier DNA

- o Another source of DNA contains the desired sequence
- o Both DNA molecules digested with the same enzyme(s)
  - Both have the same "sticky ends"
- The digested molecules are recombined
  - o Complementary base-pairing
  - o DNA ligase seals the nick in the backbone

Cloning Vectors

- Plasmids that are autonomously replicated
  - o Separate from chromosome
  - o Has its own origin of replication
- Usually carry nonessential genes
- Usually carry selectable marker
  - o Most often antibiotic resistance gene (e.g. ampR)
- Often have MCS or multiple cloning site
  - o Multiple restriction recognition sites

DNA Cloning

- Recombinant plasmid is transformed into host
  - o Not very efficient (methods don't result in 100% of the cells to be transformed)
  - o Need way to select for cells that contain plasmid
- Purpose of selectable marker
  - o Confers antibiotic resistance
  - o Grow cells in presence of antibiotic
  - o Only cells that can express the antibiotic resistance gene will survive and they are also the cells that contain the plasmid

CHAPTER 21: TRANSCRIPTION AND RNA

Transcription

- The DNA strand that guides synthesis is the "noncoding" or template strand

- The complementary strand of DNA is the coding strand

- The mRNA product is an RNA copy of the coding strand of the DNA

  - Uracil in place of thymine

Promoters

Promoters are regions of DNA that initiate transcription of genes.

RNA Polymerase, Sigma Factor (Prokaryotes)

- Core enzyme

  - α, α, β, β', ω

- Holoenzyme

  - Core enzyme + sigma (σ) factor

- Sigma (σ) Factor

  - Smaller protein

  - Guides RNA Polymerase to target DNA sequence

  - Binds to Promoter

    - -10 region (TATA box) & -35 region

  - First level of transcriptional control

Eukaryotic Promoters

- More complex promoters

- Promoter elements

  - TATA: similar to prokaryotic TATA box

  - BRE: TFIIB recognition element

  - Inr: initiator sequence; contains transcription start site

- MTE: motif ten element
- DPE: downstream promoter element
- Not all promoters contain all of these
- Work synergistically
- Transcription Factors
  - General transcription factors (6)
  - TFIIA, TFIIB, TFIID, TFIIE, TFIIF, TFIIH
    - II: specific for RNA polymerase II
  - Bind to different elements in promoters
    - TFIIB binds to BRE; TBP (of TFIID) binds to TATA (TATA Binding Protein)
  - Not a simple core complex to transcribe every gene
    - Variable subunit composition: mix and match

Transcription Regulation

- Transcription Factors – proteins involved in the process of transcription
  - TBP introduces kinks in the DNA
    - Phe side chains wedge between T and A in TATA box
    - May help other factors bind and RNA polymerase bind
  - Assembly of transcription factors produce open structure of transcription bubble

Other Regulatory Elements

- Enhancers
  - Range from 50-1500 bp in length
    - May be up to 120 kb upstream or downstream of promoter
  - Bound by activators
- Silencers
  - Bound by repressor proteins
  - May also be connected to transcription machinery via mediator

- Mediator
  - Complex of up to 60 proteins
  - Different combinations may produce different results
    - Recognize different activators/repressors
  - Not a DNA-binding complex
- Variability of gene expression
  - Prokaryotes: 1000-fold variation between most and least expressed
  - Eukaryotes: May vary by as much as 10^9-fold

Replication, Transcription, and Translation

- RNA vs. DNA

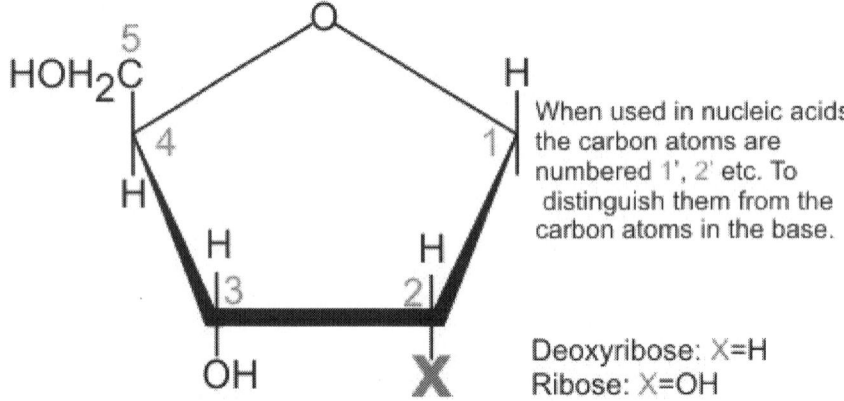

When used in nucleic acids the carbon atoms are numbered 1', 2' etc. To distinguish them from the carbon atoms in the base.

Deoxyribose: X=H
Ribose: X=OH

Deoxyribose & Ribose Sugars

  - Position of 2'-OH of ribose prevents formation of classic Watson-Crick β-helix in RNA due to steric hindrance
  - 2'-O atom would come too close to 3 atoms of the adjoining phosphate & 1 atom in next base
- What is a gene?
  - mRNA transcripts → sequence of amino acids
    - But rRNA and tRNA (& others) are not translated into protein
  - In most cases, 1 mRNA = 1 polypeptide
    - Many mRNAs in prokaryotes are polycistronic
      - Contain multiple genes

- Some mRNAs contain code for two proteins in overlapping sequences
  - DNA control elements ≠ RNA
    - These are promoters and other regulatory regions
  - RNA transcripts undergo processing before reaching functional form
    - Splicing of introns in mRNA
      - A typical gene consists of eight exons (protein-coding segments)
    - Other RNAs are processed as well
      - rRNA, tRNA
  - Noncoding RNAs
    - 2 types of genes
      - Protein-coding (DNA → RNA → protein)
      - Noncoding RNAs (DNA → RNA)
    - ~80% of the human genome may undergo transcription to produce noncoding RNAs

The lac Operon

The lac operon is required for the transport and metabolism of lactose in E.coli. An operon is a functioning unit of DNA that contains multiple genes transcribed from one promoter (operator region). All genes are transcribed together.

- Operon encodes for polycistronic mRNA
  - Contains the coding sequence for two or more genes
- Operon consists of a promoter, terminator, genes, operator
- DNA elements of the *lac* operon
  - Promoter region
    - Binds RNA polymerase
  - Operator region
    - Binds *lac* repressor protein

- o CAP site
  - ▪ Binds CAP (catabolite activator protein)
- Genes encoded by the *lac* operon
  - o *lacZ* gene encodes β-galactosidase
    - ▪ Hydrolyzes lactose to galactose + glucose
    - ▪ Converts lactose into allolactose
  - o *lacY* gene encodes lactose permease
    - ▪ Transporter for lactose and its analogues
  - o *lacA* gene encodes thiogalactoside transacetylase
    - ▪ Function is unclear
- *lacI* gene encodes the *lac* repressor
  - o Not part of *lac* operon
  - o Blocks transcription
  - o Repressor binds to operator
    - ▪ Blocks σ factor from binding promoter
  - o Always present
    - ▪ Default expression is OFF for the *lac* operon
- *lac* operon can be regulated by:
  - o Repressor protein
    - ▪ Inducible, negative control mechanism
    - ▪ Allolactose is the inducer
      - • Binds to the repressor and inactivates it
  - o Activator protein
    - ▪ Inducible, positive control mechanism
    - ▪ Small molecule involved is cAMP (cyclic AMP)
    - ▪ cAMP binds to CAP forming the cAMP-CAP complex
      - • Complex binds to the CAP site and increases transcription

- When glucose levels are high in the cell, cAMP levels are low
  - Means cAMP is not available to bind CAP
    - Transcription rate is decreased
- Mutants
  - *I⁻* mutant – has a repressor that can't bind to the operator
    - *lac* operon is constitutively expressed
  - *Iˢ* mutant – has a repressor that can't be inactivated
    - No expression
  - *O^C* mutant – has an operator that the repressor can't bind to
    - *lac* operon is constitutively expressed
  - *lacP⁻* mutant – RNA polymerase can't bind to the DNA
    - No expression

The *lac* Operon and its Control Elements

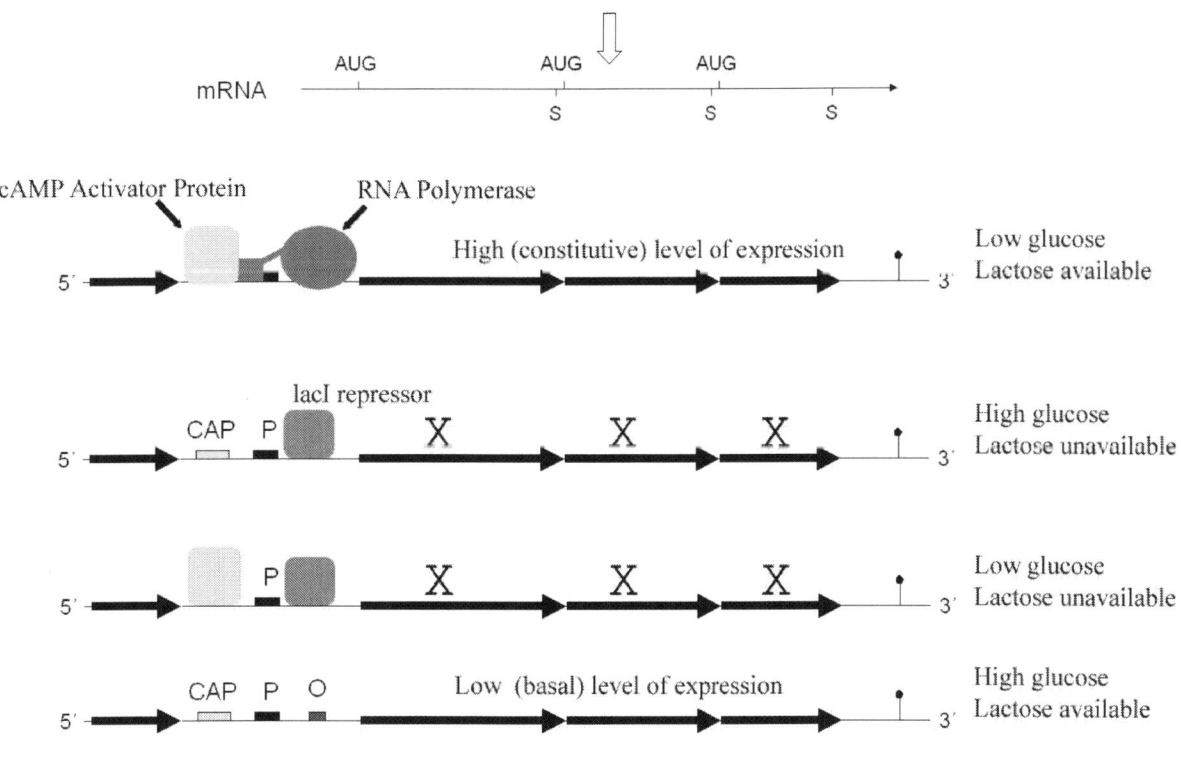

RNA Polymerase Structure

- Types of Polymerases in Eukaryotes:

    - RNA polymerase I ⇒ makes rRNA

    - RNA polymerase II ⇒ makes mRNA

    - RNA polymerase III ⇒ makes tRNA and others

- The more complex the organism, the more complex the polymerase & the promoter

- Core structure and catalytic mechanism is similar between prokaryotes and eukaryotes

- Active site is at the bottom of a positive charged cleft

- RNA Polymerase is large molecular machine

    - Binds double stranded DNA and reads the sequence

        - Reads 3' to 5'

        - Synthesizes RNA 5' to 3'

        - Does **not** need primer to begin synthesis

        - Only reads one strand (template or antisense strand); synthesizes complementary strand (coding or sense strand)

        - Substrates are rNTPs

RNA Polymerase Mechanism

- Binding the substrates

    - Incoming rNTP hydrogen bonds with the template

    - Correct rNTP causes loop of protein to close over it

        - Enhances the accuracy of transcription

    - Nucleophilic attack of 3' OH on 5' phosphate of incoming NTP

    - Synthesized in 5' to 3' direction

- Catalysis

  - Form RNA:DNA hybrid helix

  - Double helix of DNA-RNA is wider and flatter

    - Makes it less stable and easier to dissociate

  - This type of hybrid is essential in the process of synthesizing mRNA (transcription)

- Keeping things moving

  - Helix near active site (bridge) oscillates between straight and bent conformation

  - Ratchets the template to add the next nucleotide

- Releasing the product

  - "Rudder" protein loop may help separate RNA from DNA so that the RNA can exit

- RNA polymerase has proofreading ability

  - RNA:DNA helix is distorted if the wrong nucleotide is added

  - Synthesis stops and RNA backs out of active site

  - Trimmed by nuclease activity of TFIIS

Transcription Elongation

- RNA polymerase appears to abort transcription to produce short transcripts (about 12 nucleotides in length)

  - Suggests a transition from initiation to elongation

- TFIIB occupies part of the active site (blocks progress)

  - Must be displaced in order to proceed with longer transcript

- Exit channel for RNA also partly blocked

- Transition from initiation to elongation

  - Involves CTD (C-terminal domain) of RNA polymerase

  - Contains 52 repeats of a 7-amino acid sequence

    - Tyr-Ser-Pro-Thr-Ser-Pro-Ser

- o Ser residues are phosphorylated

- o No longer binds Mediator complex

- o Clears promoter and leaves general transcription factors behind

- o These can now initiate another round of transcription with a new RNA polymerase

Transcription Termination

There are two mechanisms in prokaryotes.

- Rho-independent

  - o Termination signal is in the sequence of the RNA transcript itself

  - o Hairpin forms followed by stretch of U residues

  - o Stem loop causes RNA polymerase to pause

  - o A:U base pairs are more unstable and the complex dissociates

- Rho-dependent

  - o Rho protein binds to mRNA and separates RNA from DNA (helicase)

  - o In both cases RNA polymerase moves ahead; RNA:DNA destabilized; RNA dissociates

Processing Eukaryotic mRNA

- Transcription takes place in the nucleus

  - o mRNA is processed before transport to cytosol for translation

- Processing of mRNA is closely linked to transcription

- 5' cap

  - o Protects mRNA from exonuclease digestion

  - o Phosphatase removes terminal (γ) phosphate

  - o Guanylyltransferase adds GMP from GTP

  - o Methyltransferases add methyl groups

- 3' tail

  - o Sequences in the transcript (AAUAAA) signal for cleavage

  - o Poly(A) polymerase generates a polyadenylate tail (~200 A residues)

- Poly(A) binding protein binds to help protect this end from nuclease digestion
- Half-life of mRNA depends on how rapidly the poly(A) tail is shortened
  - Deadenylating exonucleases remove protective poly(A) tail
  - Other enzymes clip off the 5' cap
  - mRNA is then digested from both ends
- Splicing
  - Eukaryotic genes contain both exons (expressed regions) and introns (intervening regions)
  - Splicing reactions cut out the introns and connect the exons
    - Starts even before transcription is finished
  - Complex of 5 small RNA molecules and 100's of proteins
  - These are small nuclear RNAs (snRNA)
  - Spliceosome
    - Recognizes specific sequences at intron/exon junctions & branch point A
      - Base pairing between mRNA and snRNA
- Splicing Reaction
  - 2' OH of branch point A residue attacks the phosphate at the 5'end of intron
    - Forms lariat intermediate
  - Free 3' OH now attacks 5' phosphate of 2nd exon
    - Intron released
  - Active site Mg_2^+ ion is essential
    - Makes -OH more nucleophilic
    - Stabilizes leaving group phosphate
- Benefits of Splicing
  - You can splice a DNA sequence in different ways
  - One gene allows for the production of multiple products

RNA Interference (RNAi)

Another way of regulating gene expression.

- Short interfering RNAs (siRNAs)

    o Double stranded RNA forms (intramolecular hairpin)

    o Dicer ribonuclease clips it to produce segments with 2-nucleotide overhang at 3' ends (siRNA)

    o siRNAs bind to RNA-Induced Silencing Complex (RISC)

    o 1 strand removed by helicase and digested (called the passenger strand)

    o Remaining strand is the guide strand

    o Guide strand directs RISC to mRNA with complementary sequence

    o Slicer of RISC cleaves mRNA and inactivates it

        ▪ Very specific

Versatility of RNA

RNA has properties that are similar to proteins.

- RNA can catalyze reactions

- RNA can adopt complex secondary and tertiary structures

- RNA nucleotides can base pair in a variety of ways

    o Watson-Crick base pairing

    o Nonstandard base pairing

CHAPTER 22: TRANSLATION

Overview of Translation

Translation is the process of translating the language of RNA into the language of proteins.

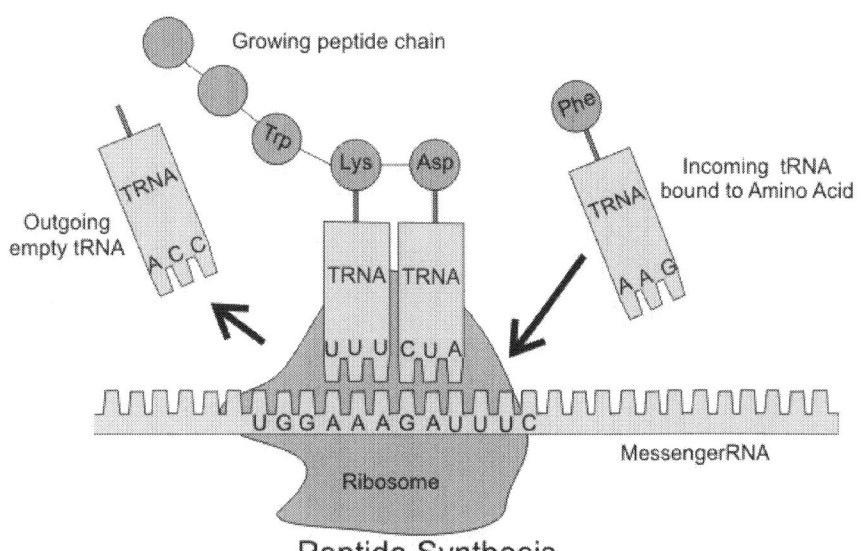

Peptide Synthesis

- Components

    o mRNA transcript - genetic code that is being translated

    o Ribosome - makes the protein

    o tRNA - amino acid carrier

    o Amino acids - what constitutes a protein

- Translation

    o 3 types of RNA participate in translation

        ▪ mRNA (messenger RNA): contains the genetic code to specify the amino acid sequence of the protein

        ▪ tRNA (transfer RNA): carries amino acids and reads the code

            • Anticodon pairs with codon

        ▪ rRNA (ribosomal RNA): part of the ribosome that makes the protein

tRNA

Set of tRNAs bind individual amino acids.

- tRNAs have specific 2-D & 3-D shape

- tRNAs have 3-base anticodon

  - Base pairs to codons in mRNA

  - Codon in 5' to 3' direction

  - Anticodon in 3' to 5' direction

- 30-40 different tRNAs in bacterial cells; up to 150 in mammalian cells

- Isoacceptor tRNAs

  - Different anticodons for the same amino acid - isoacceptor tRNAs can bind multiple codons

  - Wobble hypothesis

    - Hydrogen bonding of 3rd base in codon "wobbles"

    - The wobble hypothesis explains why many codons can be bound with fewer tRNAs

- Charging the tRNA

  - Aminoacyl-tRNA synthetase (AARS) add amino acids to tRNA

    - Requires energy from ATP

      - AMP added to amino acid

    - Amino acid attached to tRNA at acceptor arm

      - Connected at α-COO- group

      - α-NH_2 group free

    - Ester bond formed with 3'-OH of tRNA

    - Most cells contain 20 AARS's

      - Isoacceptor tRNAs recognized by same AARS

      - Same amino acid attached

- o Aminoacyl-tRNA synthetase reaction

  - ▪ Amino acid "charged" by addition of AMP from ATP

  - ▪ Subsequent hydrolysis of pyrophosphate (PPᵢ) makes this irreversible

  - ▪ Amino acid transferred to tRNA and AMP released

- o Specificity of AARS

  - ▪ AARS protein must recognize 2 of tRNA

    - • The contact anticodon loop and the acceptor arm

    - • Ensures correct amino acid is connected to its cognate tRNA

  - ▪ The AARS may also have proofreading ability: **example of Ilv eRS**

    - • 1st active site excludes any hydrophobic amino acid larger than Ile

    - • Ile and other smaller amino acids could get activated

    - • 2nd active site hydrolyzes any smaller than Ile (Ile won't fit in this pocket)

Ribosome Structure

The ribosome has two major components, the small ribosomal subunit which actually reads the RNA and the large subunit which joins the amino acids together to form a polypeptide chain. Each of the subunits is composed of rRNA and a variety of proteins.

- • 2 subunits, 52 proteins, 3 rRNAs

- • Can bind 1 mRNA + 3 tRNAs

- • Binding sites for tRNA

  - o A site: carries tRNA with the next amino acid to be added

    - ▪ Acceptor for the growing protein during peptide bond formation

  - o P site: holds the growing peptide chain

  - o E site (exit): discharged tRNAs leave the ribosome here

- • mRNA makes a sharp bend between codons in A and P sites

  - o This allows 2 tRNA molecules to fit side-by-side

  - o May also prevent the ribosome from slipping

Prokaryotic Translation Initiation

- Initiator tRNA

    - Methionine is the initiator tRNA as derivatized with an N-formyl group

    - The initiation codon is recognized by an initiator tRNA that has been charged with methionine

- Initiation in Prokaryotes

    - Small ribosomal subunit binds to Shine-Dalgarno sequence

        - 16S rRNA base pairs with this sequence in the mRNA

        - Looks for 1st AUG codon

    - Three initiation factors are required for translation

        - mRNA and fMet-tRNA in complex with IF-2-GTP bind to the small (30S) ribosomal subunit

            - IF-1 binds to A site to block entry there

            - IF-3 prevents 50S from associating until tRNA is in place

        - Association of the large (50S) subunit with the 30S subunit triggers IF-2 to hydrolyze its bound GTP

        - The tRNA bearing the initial fMet is positioned in the P site of the ribosome

Eukaryotic Translation Initiation

- Initiation in eukaryotes

    - 12 initiation factors (eIF's)

        - Recognize 5' cap and poly(A) tail

        - Small ribosomal subunit scans to find the 1st AUG

    - Eukaryotic initiation factors of translation are often G-proteins

    - GTP hydrolysis creates conformation changes

Translation Elongation

- Aminoacyl-tRNAs and EF-Tu

    - EF-Tu: prokaryotic elongation factor

- o Aminoacyl-tRNAs are delivered to the ribosome in complex with an elongation factor
- o EF-Tu won't bind uncharged tRNA
- o All amino acid carrying tRNAs bind EF-Tu with roughly the same affinity
- Ribosome Helps Codon-Anticodon Pairing
  - o 30S subunit confirms that the pairing is correct
  - o 2 alanine residues of 16S rRNA flip out to make contact with mRNA
    - Contacts not made if pairing between tRNA and mRNA is off
    - Rate-limiting step of protein synthesis
- EF-Tu as a Proofreader
  - o EF-Tu-GTP delivers an aminoacyl-tRNA to the A site of the ribosome
  - o If the tRNA anticodon matches the mRNA codon, Ef-Tu hydrolyzes its GTP and dissociates from the ribosome, leaving the aminoacyl-tRNA in the A site
    - Ribosome ready is ready for transpeptidation
  - o If the tRNA anticodon and mRNA codon are mismatched, the aminoacyl-tRNA dissociates before the Ef-Tu hydrolyzes GTP

Transpeptidation

- Free amino group of aminoacyl-tRNA in A site attacks the ester bond of the peptidyl group in the P site
  - o Results in the growing peptide chain being lengthened by 1 amino acid
- Movement is from P to A site
- This is why polypeptide chain grows from N to C terminus
- This is an example of a ribozyme
  - o RNA catalyst

Protein Folding

- Molecular Chaperones
  - o Use the energy of ATP hydrolysis to help the protein refold

- Trigger Factor
    - Positioned just outside the exit tunnel of the ribosome
    - Binds to hydrophobic patches of protein
    - May hand off the protein to another chaperone
- DnaK
    - Clamps down on protein then releases it to allow it to fold
    - Binds to hydrophobic patches
- GroEL/ES
    - 2 GroEL heptameric rings
        - Protein folding occurs inside the rings
        - Each subunit binds and hydrolyzes 1 ATP (7 per ring)
    - Binding of GroES "cap" triggers release of peptide into GroEL chamber
    - Doesn't actually fold the protein itself
        - Just provides a protected environment for the protein to fold itself

Protein Translocation

Protein translocation is the process by which peptides are transported across a membrane bilayer.

- Some polypeptides carry a signal peptide
    - Short α helix (hydrophobic) preceded by a positively charged residue
- This is the signal to translocate the protein
- Bound by the SRP (signal recognition particle)
- Has pocket of mainly Met residues
- Also has segment of negatively charged RNA that interacts with the positively charged residue
- Once translocated, the signal may be removed

Post-Translational Modifications

- Many proteins are modified after translation

- Proteolysis – breakdown of proteins into smaller polypeptides
- Groups can be added to side chains
  - Methyl, acetyl, propionyl
- Glycosylation
- Lipid anchors or fatty acyl chain may be added

- Proteins may vary considerably, beyond just the amino acid sequence indicated in the genetic code

CONCLUDING REMARKS

I hope this book has provided you tremendous value for your money and has helped you do better on your exams! If it has done both of these things, I have achieved my purpose in making this guide.

Furthermore, my goal is to create more books and guides that continue to deliver great value to readers like you for little monetary costs. Thank you again for purchasing this study guide and I wish you the best on your future endeavors!

- Dr. Holden Hemsworth

More Books By Holden Hemsworth

Do You Need Help with Other Classes?

Check out Other Books in the Ace! Series

 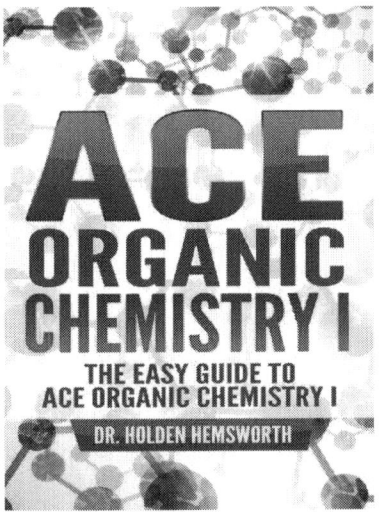 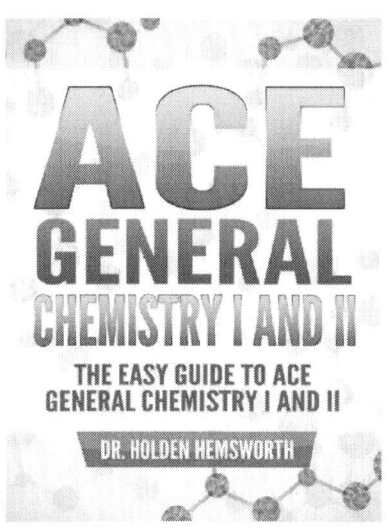

  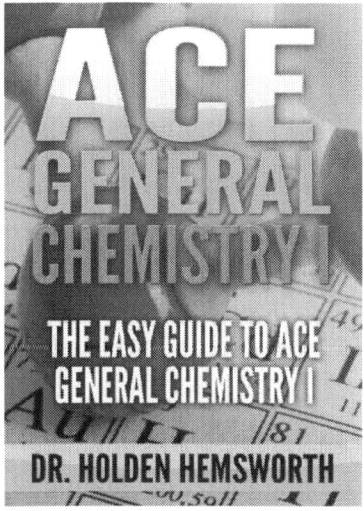

All Books are Listed on My Amazon Author Page

More Books Coming Soon!

Printed in Great Britain
by Amazon